SELECTED WORKS 2012-2013

中国建筑设计研究院　　China Architecture Design & Research Group
中国北京车公庄大街19号　　No.19 Che Gong Zhuang St.
　　　　邮编：100044　　Beijing 100044, P. R. China
电话：010-88328888 68348613　　Tel:86-10-88328888 68348613
传真：010-88373837 68348832　　Fax:86-10-88373837 68348832
　　电子信箱：yy@cadg.cn　　E-mail:yy@cadg.cn

作品 2012-2013
中国建筑设计研究院作品选
SELECTED WORKS 2012-2013
OF CHINA ARCHITECTURE
DESIGN & RESEARCH GROUP

中国建筑工业出版社
China Architecture & Building Press

前言
PREFACE

这是中国建筑设计研究院连续出版的第七部作品选，收录了2012-2013年已经落成的部分项目，集中展示了我院设计领域的新成就。在这些值得称道的作品中，我们也许可以从中体味到一些"本土设计"理论所赋予的精髓。诚然，近年来我院的设计师在这个方向上进行了不断的探索与追求，并且逐步形成共识。

恰逢我院郑世伟建筑师主编的《建筑·十四室记》一书出版，书中写道"我们是一个国有大型建筑设计集团里面的一个小建筑工作室……我们一直在思考，在这个国有大院里的小设计团队的存在价值和未来设计的方向。我们希望以一个积极的社会角色去调和权利、资本与公众利益之间的矛盾，营造更多足够吸引人参与的公共场所、它们有人文关怀，能引导大家自由地参与各种真实、生动的社会活动，感受自我在社会中的存在……"青年人的所思所想，已经超越了传统建筑设计偏重技法的一面，而是站在了作为主流设计院的一员如何对待建筑设计的高度上。本册作品选也收录了他们设计的"苏州市吴江中学"等项目。一幅幅精美的图片，更能折射出一代青年建筑师、工程师的璀璨亮点和强烈的社会责任感。

每一个设计项目都是一个"梦"。作为设计师这个职业值得庆幸的是，我们的"梦"大多是可以实现的。正是因为这些梦想的实现性，也就增加了对于自身职业的敬畏感。伴随我国城镇化进程的加快，"望得见山，看得见水，记得住乡愁……"将成为每一位建筑师、工程师立足本土，不断进取的方向。

期待第八部作品选更加精彩。

刘燕辉

中国建筑设计研究院　副总建筑师

As the seventh Selection of Works from the China Architecture Design & Research Group, this book exhibits our latest achievements in completed projects from 2012 to 2013. Senses of native design can be appreciated from these projects, more or less, as designers from the institution have been constantly exploring in this direction and have achieved some consensus all these years.

It happens that a book, *Architecture, Stories of 14th Studio*, by Mr. Zheng Shiwei from our institute is published."As a small studio in a large architecture design group…we have been thinking about this: what is the value of our existence and what is the future way of design. We hope that, as a positive social role, we can take part in reconciling contradictions between power, capital and public interest, so that more attractive spaces can be created where people can feel humanistic care and can join all kinds of social activities to prove their existence in this society…"Thoughts from young architects have already surpassed the traditional ones that emphasized on techniques only, instead, they, as members of the mainstream design institutions, are thinking about how to value the architecture design itself. In this book, there are some projects from them, such as Suzhou Wujiang Middle School and so on. With these beautiful pictures, one can feel shining merits and strong social responsibilities from these young architects and engineers.

Each design project is a dream as well, it is lucky that, as designers, we can realize most of our dreams. Yet due to the possibilities of realization of these dreams, we also need to show more respect to our career. Along with the rapid progress of urbanization, it is important for every designer to stick to our homeland before making any progresses; we should always remember the hills, the rivers, and the nostalgia…

Let's look forward to the next book of fantastic selected works.

Liu Yanhui

Vice Chief Architect of CADREG

目录 CONTENTS

08　重庆国泰艺术中心
　　Chongqing Guotai Arts Center

16　绩溪博物馆
　　Jixi Museum

24　中信金陵酒店
　　CITIC Jinling Hotel

30　承德行宫大酒店
　　Chengde Imperial Palace Hotel

36　中信泰富朱家角锦江酒店
　　CITIC Pacific Zhujiajiao Jin Jiang Hotel

44　清远狮子湖喜来登度假酒店
　　Sheraton Qingyuan Lion Lake Resort

52　益田影人花园酒店
　　Cineaste Garden Hotel

56　中华人民共和国驻南非大使馆
　　The Embassy of the People's Republic of China in the Republic of South Africa

64　中国驻开普敦总领事馆
　　The Consulate General of the People's Republic of China in Cape Town

68　新城大厦二期
　　Xincheng Building, Phase II

72　南昌联发广场
　　Nanchang Lianfa Plaza

74　江西艺术中心
　　Jiangxi Arts Center

78　昆山市文化艺术中心
　　Kunshan Cultural Arts Center

86　德州大剧院
　　Dezhou Grand Theater

92　青海艺术中心
　　Qinghai Arts Center

96　大沽口炮台遗址博物馆
　　Dagukou Fort Ruins Museum

99　蓬莱古船博物馆
　　Penglai Ancient Ship Museum

104　中国杭帮菜博物馆
　　Chinese Hangzhou Cuisine Museum

110　濮阳市城乡规划展览馆
　　Puyang Planning Exhibition Hall

112　宁波国际贸易展览中心2号馆
　　Ningbo International Trade & Exhibition Center No.2 Pavilion

114　青海科技馆
　　Qinghai Science & Technology Museum

116　鄂尔多斯机场新航站楼
　　New Terminal of Erdos Airport

124　大同机场新航站楼
　　Datong Airport New Terminal

128　苏州火车站站房
　　The Building of Suzhou Railway Station

134　苏州市吴江中学
　　Suzhou Wujiang Middle School

140　南京艺术学院改扩建项目
　　Renovation of Nanjing Arts Institute Campus

152　北京外国语大学图书馆新馆
　　New Library of Beijing Foreign Studies University

156　北京华文学院新校区
　　New Campus of Beijing Chinese Language and Culture College

160　德阳市奥林匹克后备人才学校
　　Deyang Olympic Sports School

166　鄂尔多斯东胜图书馆
　　Erdos Dongsheng Library

170	万州体育中心 Wanzhou Sports Center	218	北京风景 Beijing Scenery
174	昌吉体育馆 Changji Stadium	222	远洋一方二期 Poetry of River, Phase II
180	北工大软件园二期E地块 BPU Software Park II-Plot E	226	万科·蓝山 Vanke Beijing Hills
184	中国科学院电子学研究所怀柔园区 Huairou Park of Electronics Institute, Chinese Academy of Sciences	230	中海·苏黎世家 China Overseas·Zurich Garden
188	绿色建筑材料国家重点实验室 State Key Laboratory of Green Building Materials	232	太湖湾云顶天海度假村 Taihuwan Yunding Tianhai Village
190	巴彦淖尔市临河区行政中心 Administrative Service Center of Linhe District, Bayan Nur City	234	兰州鸿运润园集成示范住宅 Integration Demonstrative Residence of Hongyun Run Garden, Lanzhou
194	巴彦淖尔市政务服务审批中心 Administrative Approval Service Center of Bayan Nur City	238	复地朗香别墅四期 Forte Ronchamps Villa, Phase IV
198	镇江新区金融大厦 Financial Building of Zhenjiang New District	240	南京苏宁睿城 Nanjing Suning Smart City
202	宁波市交通运输委员会办公楼 Office Building of Transport Committee of Ningbo Municipality	242	大同市妇女儿童医院 Datong Women's & Children's Hospital
204	宁波市工商局办公楼 Office Building of Ningbo Administration for Industry & Commerce	244	大同市中医院 Datong Chinese Medicine Hospital
206	玉树康巴风情商街 Yushu Khamba Style Commercial Street	246	山西省肿瘤医院放疗医技综合楼 Comprehensive Building of Shanxi Provincial Tumor Hospital
210	常州方圆·云山诗意 Changzhou Fineland·Orient Residential Area	248	欧美同学会改扩建 Extension of Western Returned Scholars Association
214	四合上院 Sihe Up Courtyard	252	威克多制衣中心 Vicutu Garments Manufacturing Center
216	重庆线外SOHO及会所 The Forward - SOHO & Club, Chongqing	256	中间建筑A、F区 A/F Plots of the Inside-out
		262	八达岭太阳能热发电站吸热塔 Thermal Absorbtion Tower of Badaling Solar Thermal Power Station

重庆国泰艺术中心 Chongqing Guotai Arts Center

地点 重庆市渝中区 / **用地面积** 9 600m² / **建筑面积** 36 170m² / **高度** 51m / **设计时间** 2005年 / **建成时间** 2013年

方案设计	崔愷 景泉 张小雷 马志新
设计主持	崔愷 秦莹 景泉
建 筑	李静威 张小雷 杜滨 邵楠
结 构	张淮湧 施泓
给排水	靳晓红
设 备	孙淑萍
电 气	梁华梅 许士骅
总 图	余晓东
室 内	张晔 刘烨 饶劢
景 观	赵文斌

国泰艺术中心的色彩组合既鲜明又与传统文化有所契合，在林立高楼的间隙间，露出红色的边角，为人们提供了方向感和归属感。

Chongqing Guotai Arts Center has a distinctive regional characteristic in colors, with a sense of direction and belonging by revealing red corners in between tall buildings.

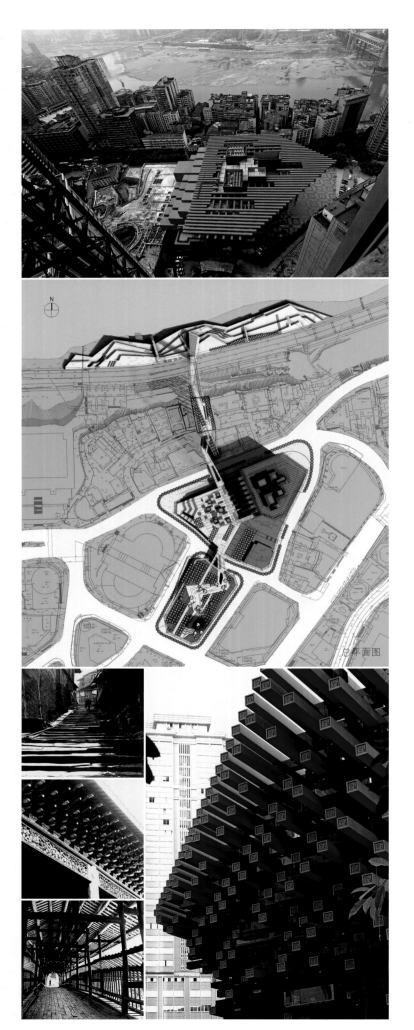

总平面图

国泰艺术中心位于重庆市CBD核心区，作为标志性建筑对其他地块形成统领作用，其形象既统一于解放碑地区现有建筑，又为该地区创造了新的秩序。建筑结合自身功能的复杂性，形成与外部城市空间相互融合的肌理。单纯的筒状构件，在相互穿插、叠落、悬挑中，产生联系、渗透，因而生成建筑的整体形象。这种从简单累加而成的复杂，与普遍存在于汉代的一种营造工法"题凑"有所契合。其中的"题"指木头的端头，"凑"指排列的方式。建筑也因此以简洁而现代的手法体现了传统建筑的内涵。其高高迎举、顺势自然的状态，也正是重庆人最本质的精神追求。

Chongqing Guotai Arts Center occupies a pivotal site within the CBD of Chongqing city. Intensively surrounded by tall buildings, it plays a commanding role in the formation of other plots as a landmark. In harmony with the existing buildings it creates a new order in the local commercial center – Jiefangbei area. It receives a texture featured with oriental characteristics in mutual penetration with external urban spaces by combination of the complexity of its own architectural functions. By the stacked, intersected and cantilevered sticks, the building forms its own overall image of infiltration and connection of different textures. The stick system could be recognized as an inheritor of the traditional construction method "Ti-cou" which was common in Han Dynasty. "Ti" refers to the ends of the wood, and "cou" is the construction method. Therefore, such a concise and modern architectural concept represents the value of traditional Chinese building. The active and courageous feature also reflects the most essential spirit of Chongqing people.

建筑造型最初的构思来源于重庆湖广会馆中一个多重斗栱构件，利用传统斗栱的空间穿插形式，以现代简洁的手法表达传统建筑的内涵。

The architectural form derives from a multi-component bracket set in Huguang Guild Hall in Chongqing, with traditional brackets space interspersed. It expresses the spiritual content of traditional architecture in a modern way.

处于建筑外部的红色构件是建筑通风系统，黑色构件则内蓄水作为冷媒，两套构件共同发挥作用形成建筑外部生态节能系统。红与黑的交叠错动同时也引入室内，成为室内装饰的主旋律。

The red components act as the ventilation system and the black components stored with water are the refrigerant. The stacking of red and black components is also invited inside as the theme of the interior decoration.

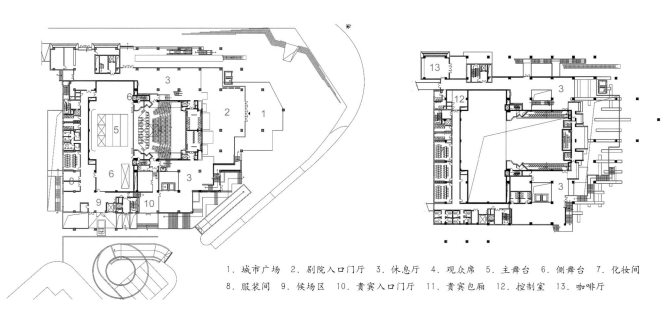

1. 城市广场 2. 剧院入口门厅 3. 休息厅 4. 观众席 5. 主舞台 6. 侧舞台 7. 化妆间
8. 服装间 9. 候场区 10. 贵宾入口门厅 11. 贵宾包厢 12. 控制室 13. 咖啡厅

−2.100标高平面图　　　　　　　　　5.700标高平面图

建筑的东西两侧及中部形成一系列吹拔空间，其中有一些平台、楼梯，成为与城市共享的灰色空间，既有利于大型建筑的人员疏散，又为城市创造了更多市民休闲空间。

In the east and west sides as well as the middle part, there is a series of patios, among which a few grey spaces are created with terraces and stairs to be shared with the city. It is conducive to the evacuation of people in large buildings, and creates more leisure space for citizens.

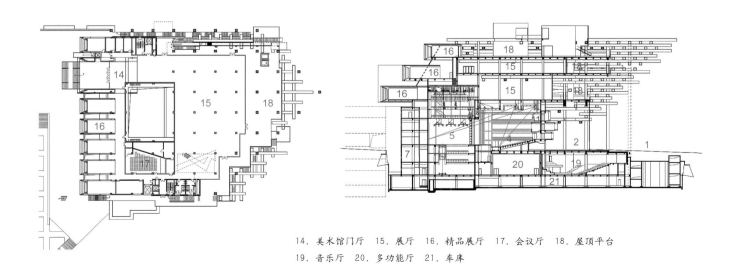

14. 美术馆门厅　15. 展厅　16. 精品展厅　17. 会议厅　18. 屋顶平台
19. 音乐厅　20. 多功能厅　21. 车库

17.850标高平面图　　　　　　　　　　　　剖面图

绩溪博物馆　Jixi Museum

地点 安徽省绩溪县 / **用地面积** 9 290m² / **建筑面积** 10 003m² / **高度** 14m / **设计时间** 2009年 / **建成时间** 2013年

方案设计　李兴钢　张音玄　张哲
　　　　　　易灵洁　张一婷　闫昱
设计主持　李兴钢　张音玄
建　　筑　张哲　邢迪
结　　构　杨威
给 排 水　杨兰兰
设　　备　李京沙
电　　气　丁志强
总　　图　王炜
景　　观　李力
摄　　影　李兴钢

绩溪博物馆的基址曾为县衙，后为县政府大院，因古城整体保护规划而修建为博物馆。建筑设计基于对绩溪的地形地势、名称由来和徽派建筑聚落的考察。整个建筑覆盖在一个连续的屋面之下，起伏的屋面轮廓仿佛绩溪周边山形水系，是对"北有乳溪，与徽溪相去一里，并流离而复合，有如绩焉"中溪水并行交错之地貌的充分演绎和展现。为尽可能保留用地内的现状树木，建筑的整体布局中设置了多个庭院、天井和街巷，既营造出舒适宜人的室内外空间环境，也是对徽派建筑空间布局的重释。建筑群落内沿街巷设置了东西两条水圳，汇聚于主入口庭院的水面。庭院正对入口设置了一组被抽象化的"假山"。围绕庭院、水系、街巷设置的立体观赏流线，带领游客在博物馆中上下俯览，遍赏丰富的空间、起伏的屋顶和秀美的远山。在适当采用当地传统建筑技术的同时，设计以灵活的方式使用砖、瓦等当地常见材料，尝试使之呈现出当代感。

The design of the museum is based on a delicate research of the topography and history of Jixi, a hilly county located at the foot of Mt. Huangshan and famous for its vernacular settlements. Because of the chains of hills, there are a number of rivers flowing along the valleys parallelly that generate the origin of the name Jixi, "brooks look like the threads of a string". A continuous canopy is occupied to cover the entire site. Its profile simulates the undulating feature of the surrounding hills and rivers. Various courts, patios and alleys are carved out within building to maintain the original trees of the site as much as possible. Two streams along the alleys join into the water surface of the entry court, where a vestibule with a set of modernized rockery in front of it forms a backdrop to court for the entering visitors. A three-dimensional touring route began with the entry is developed around the courts, steams and lanes to guide visitors to encounter the exhibition halls, undulating roofs and distant hills. Local construction materials such as bricks and tiles are used in a flexible manner to gain a contemporary appearance.

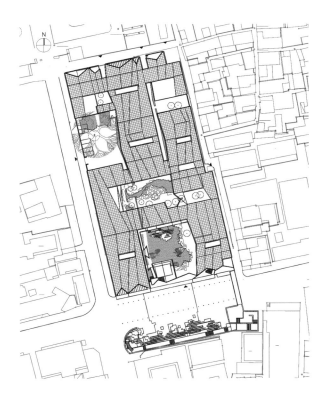

总平面图

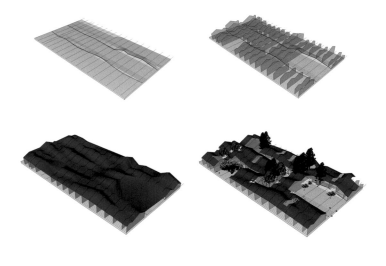

规律性组合布置的三角屋架单元,其坡度源自当地建筑,并适应连续起伏的屋面形态。部分屋顶被挖除以形成院落。

The gradient of the regularly arranged triangular roof trusses derives from the roof of local buildings. Arrayed in a random but rational manner, they feature a continuous and undulating roof, which is punctuated by a series of patios.

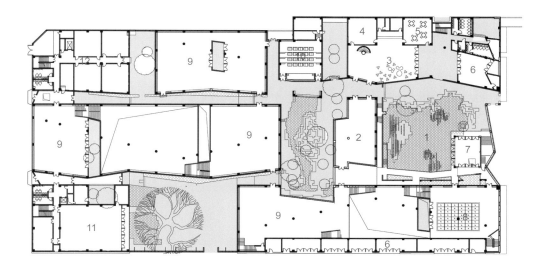

1. 前庭 2. 序言厅 3. 接待厅 4. 贵宾厅 5. 教室 6. 商店
7. 茶亭 8. 保留县街遗址 9. 展厅 10. 影院 11. 报告厅 12. 辅助用房

首层平面图

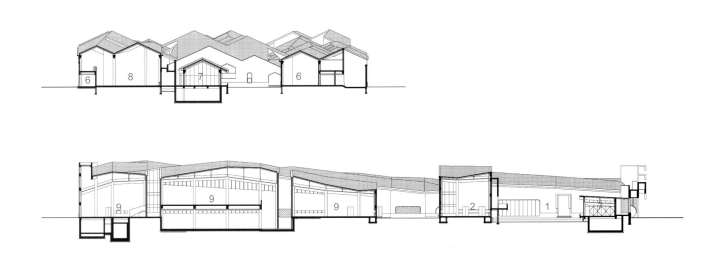

剖面图

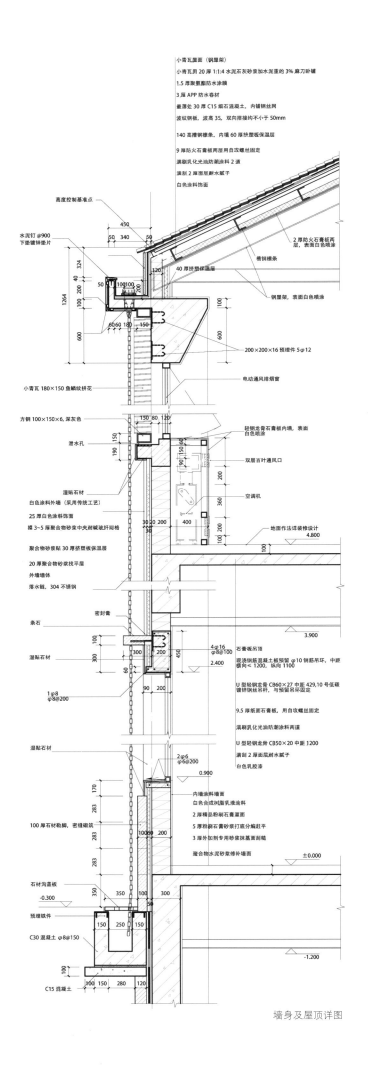

墙身及屋顶详图

中信金陵酒店　CITIC Jinling Hotel

地点 北京市平谷区 / 用地面积 251 260m² / 建筑面积 44 460m² / 高度 34m / 设计时间 2010年 / 建成时间 2012年

方案设计	崔愷 周旭梁 刘恒 潘观爱 时红 梁丰
设计主持	崔愷 时红
建　筑	周旭梁 赵晓刚 梁丰 金爽 周力坦 张汝冰
结　构	朱炳寅 王奇 宋力 杨婷 郭天晗
给排水	王耀堂
设　备	徐征 祝秀娟
电　气	许冬梅 王莉
智能化	张月珍
总　图	王雅萍
景　观	谢晓英

中信金陵酒店位于北京郊外的山坳中，西北侧正对西峪水库。作为一处郊外度假酒店，建筑以"栖山、观水、望峰、憩谷"为主题，依山就势，产生层层跌落的形态，体现出与自然环境的融合。巨大的山石作为基本形态，标志出建筑的公共空间。建筑的中心位置是五层通高的大堂，空间层层叠退，引向内部，上部插入的天窗则如钻石般，产生变幻多姿的光影效果。两侧客房顺山势台阶状跌落，充分利用了山地环境及良好的景观朝向，实现了室内外空间的有机渗透。GRC挂板、生态木和玻璃等材料的搭配运用，赋予建筑粗犷、壮美的外观。尤其是GRC挂板，其表面肌理拓自真实的山石，并加入石粉以模拟天然色彩，凹凸明显，表情粗犷。相邻块材间彼此凹凸交错，表现出垒砌的关系，充分贴近自然。

Nestled against the mountains, the hotel is designed to enjoy the beauty of the northwest reservoir and the surrounding hills. The site strategy for the destination resort establishes an intimate connection to the mountain by the stepping backwards terraces. The public spaces are defined by several huge rock-like volumes, which also create fantastic effects by punching into interior. A 5-story-high lobby, the kernel of the building, is characterized by the step back floors and crystal-like irregular skylights and flanked by two wings of guest rooms. Terraced layout is quite fit for the topography and gives every room a great view. The material palette of GRC panels, GreenerWood and glass expresses the hotel's respects to the nature. In particular, those GRC panels made by rock flour imitate the texture and color of natural rocks yet allow the structures to seemingly emerge from the land.

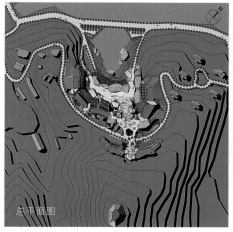

总平面图

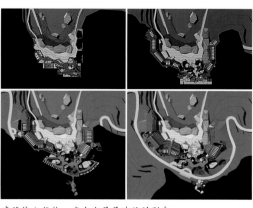

建筑依山就势，产生出层层跌落的形态
The cascading forms mimic the topography of the mountains

摄影 耿毅军

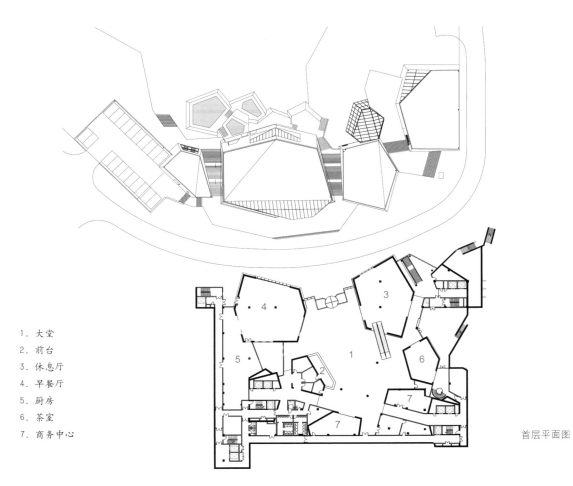

1. 大堂
2. 前台
3. 休息厅
4. 早餐厅
5. 厨房
6. 茶室
7. 商务中心

首层平面图

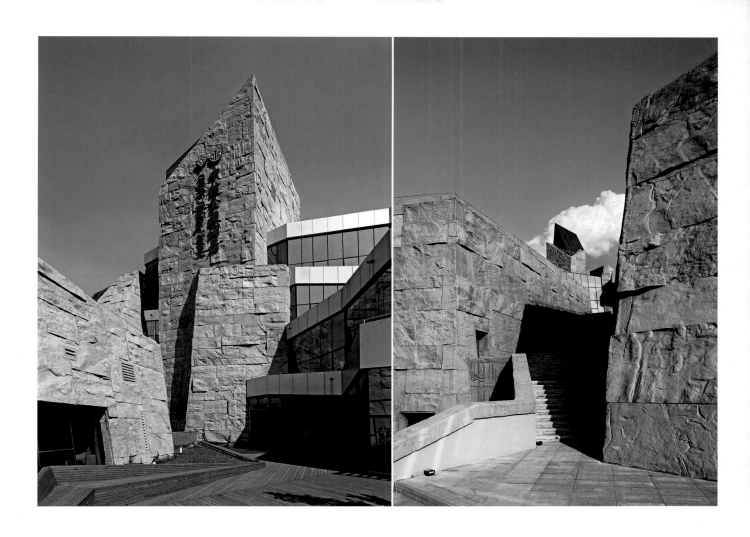

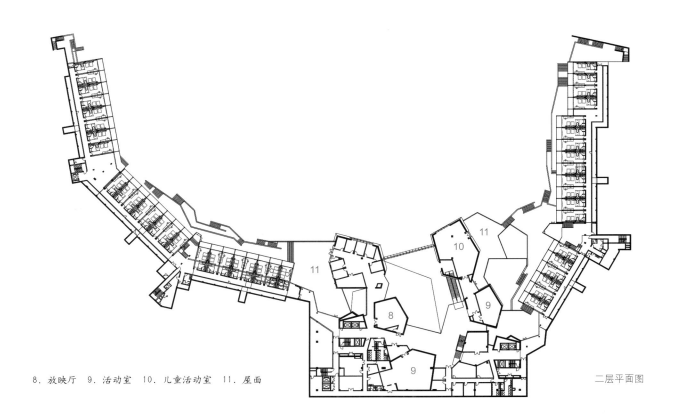

8. 放映厅　9. 活动室　10. 儿童活动室　11. 屋面

二层平面图

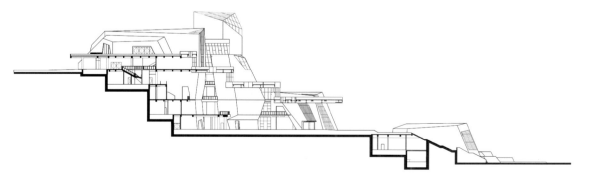

剖面图

承德行宫大酒店　Chengde Imperial Palace Hotel

地点 河北省承德市 / 用地面积 25 798m² / 建筑面积 28 093m² / 高度 14m / 设计时间 2009年 / 建成时间 2012年

方案设计	柴培根　王效鹏　杨凌
	周凯　蒋鑫
设计主持	柴培根
建　　筑	王效鹏　杨凌
结　　构	霍文营　孙洪波　郝国龙
给 排 水	赵昕
设　　备	张亚立
电　　气	陈红
电　　讯	都治强
总　　图	刘文

行宫大酒店地处承德著名的双塔山景区脚下。位于城市主干道和自然山体之间的基地，其最大进深不足200m。设计因此将注意力放在对自然的回应上，建筑以水平方式嵌入场地，并通过布局传递出从城市到自然的体验，前为严整对称的酒店公共区，后为自由分散的客房私密区，这也是借鉴避暑山庄宫苑格局中宫殿与山水对比关系的结果。五个主题院落相互串联递进，中心庭院以水为主；其他则以绿化为主，并结合北侧的山势，借景双塔山，使住客产生居于自然之中的感受。

Situated at the foot of Shuangta Hill, the historic spot of Chengde, the Imperial Palace Hotel owns a narrow site between a busy city road and the natural hill. A horizontal developed complex is the answer to resolve the site's contradictory nature. The public area facing the road is strictly-symmetrical, while the guest room area on the north which is close to the scenery occupies a looser plan. This layout could be considered as a modern translation of the juxtaposition of palace and garden in the Imperial Summer Resort of Chengde.

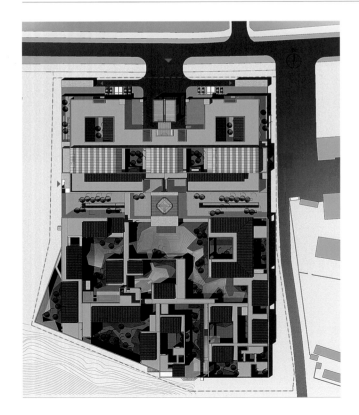

总平面图

绿化

院落空间

水系

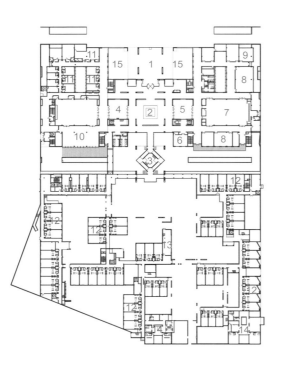

首层平面图

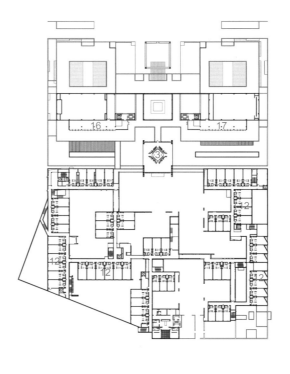

二层平面图

1. 门厅
2. 四季厅
3. 过厅
4. 前台
5. 商务中心
6. 商店
7. 大会议室
8. 会议室
9. 休息区
10. 早餐厅
11. 包间
12. 客房
13. 咖啡吧
14. 餐厅
15. 院子
16. 酒吧
17. 雪茄吧

前区包括大堂、餐厅、多功能厅、会议室等功能，强调公共区域的礼仪性，空间易于识别，院落呈独立静止状态，后区的客房结合不同院落分组布置，以产生私密性和识别性。通过院落切削的建筑形体，表现为轮廓的曲折变化，削弱了实体感，减少了与双塔山的视觉冲突。

Housing public functions such as lobby, restaurants, multi-function hall and meeting rooms, the public area is arranged to emphasis the formality and identity of spaces. The courtyards here are enclosed by building volumes and independent to each other. On the contrary, the guest room area is dominated by a series of gardens. They form a continuous landscape sequence and group the guest rooms into five clusters. As the remainder of a solid volume carved by the gardens, the mass of buildings is more irregular and less of visual conflict to the hill.

景观设计作为一个系统与建筑布局紧密结合,其中核心的元素是水。水面系统将各个院落串联起来,一方面给人行路径提供了边界及指向性,另一方面建立了院落之间的视觉联系和想象空间。层层叠涩的浅水池,采用看似随意的折线,同样体现了水的流动性和连续感,不同于传统园林中叠石设障的方式,呈现了无缝衔接的关系,使建筑和水面的对话变得更加直接。

The landscape design, as a counterpart of architecture, regards water as its soul element. By stringing all the gardens, the water system not only provides a visit route but also builds up the visual connection of different gardens. Without laying stones typically used on the waterfront of traditional garden, the zigzag of water edges visualizes the water flow and intensifies the relation between building and water surface.

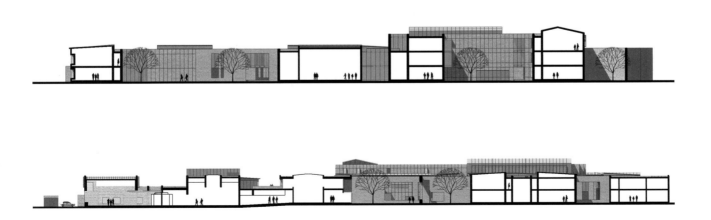

剖面图

中信泰富朱家角锦江酒店　CITIC Pacific Zhujiajiao Jin Jiang Hotel

地点 上海市青浦区 / **用地面积** 43 449m² / **建筑面积** 40 430m² / **高度** 12m / **设计时间** 2007年 / **建成时间** 2013年

方案设计	崔愷 刘恒
	何理建 李斌
设计主持	崔愷
建　　筑	单立欣 刘恒
	白海 施海燕
结　　构	贾卫平 丁卫国
给 排 水	李严 夏丽华
设　　备	郭晓楠 许敬逸
电　　气	王晋恒 宋建国
电　　讯	张月珍 宋建国
总　　图	白红卫

宾馆与朱家角古镇隔大淀湖相望，以若干大小不同的矩形院落的相互交叉嵌套，用方整的单体形成了丰富而生动的聚落式空间。格栅、木窗、片石、竹林等传统元素的运用，均衡而非对称的构图方式，使素白的现代建筑与江南水乡的传统民居产生了契合。每个院子具有独特的景观和装修形式，并结合陶瓷、竹木、琉璃、砖墙、金属等不同材料形成各自的主题。院落围合和建筑墙体上看似随意的开窗，既来自功能需求，也根据不同景观方向，形成巨幅景框，将自然风景和人文艺术剪裁成迷人的画卷。沿着建筑内部行走其间，江南园林般的空间形态次第展开，现代与传统交织其中。

Lies across Dadian lake to Zhujiajiao old town, the resort adopts a spatial organization casting a number of boxes with different sizes, proportions and textures into the site. The interweaving and interaction of modern cubes create a dramatic atmosphere with the scene of traditional settlement. Detailed by the lattice, rockery and bamboo, the whole resort is a blending of modern construction with a subtle reference to historical building form in the riverine towns. Outside, the rich manners of decoration, landscaped design and building material individualize each courtyard. The openings of courts and walls frame lots of giant views of the lake and human landscape. Inside, a rich procession of space is formed that recalls the experiential sequence of traditional Chinese gardens.

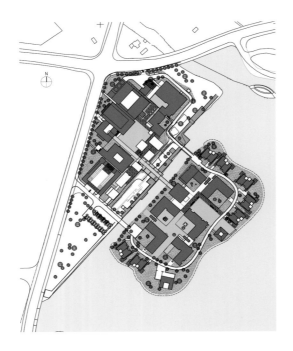

为了保持环湖公共绿带的完整，建筑基地被一分为二，其高度也因毗邻古镇而受到限制，控制在12m以下。

Divided by a green loop around the Dadian Lake, the resort near to the old town should to be designed under the strict height constraints of historical zone.

总平面图

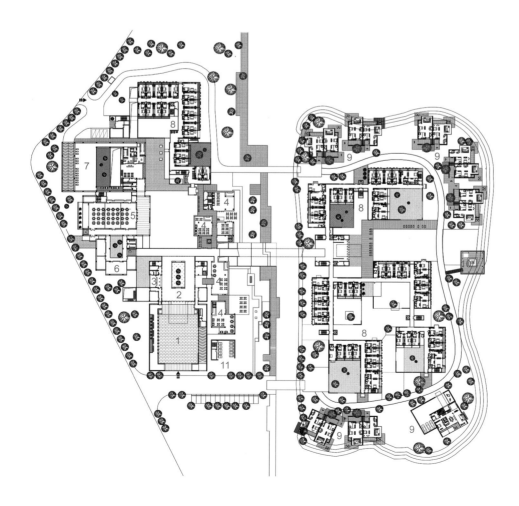

1. 入口庭院
2. 大堂
3. 总服务台
4. 餐厅
5. 宴会厅
6. 会议室
7. 游泳池
8. 客房
9. 独立式客房
10. 茶室
11. 电瓶车等候区

首层平面图

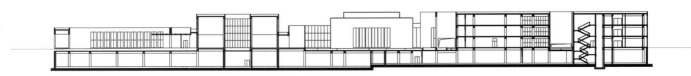

剖面图

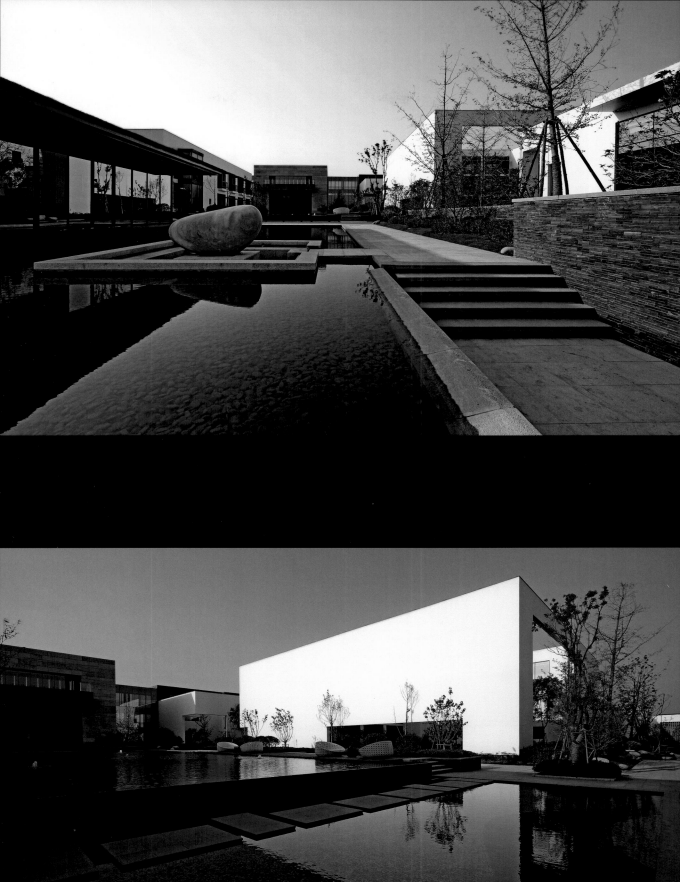

清远狮子湖喜来登度假酒店　Sheraton Qingyuan Lion Lake Resort

地点 广东省清远市 / **建筑面积** 60 000m² / **设计时间** 2008年 / **建成时间** 2011年

设计机构 深圳华森建筑与工程设计顾问有限公司
＋美国WATG建筑设计集团

清远狮子湖以高尔夫球场著称，拥有丰富的生态环境，开发商因此决定在这里兴建一处包括两座酒店、渔人码头、会所和别墅在内的度假胜地。其中的超五星级喜来登度假酒店，由两座客房楼、会议中心和SPA中心组成。酒店大堂和会议厅出入口位于7m高的花园平台上，使客人在主要活动空间中能够方便地俯瞰湖景。临湖主要布置客房和餐厅、酒吧、SPA等休闲空间，海鲜舫和SPA的VIP房更设于湖面之上，以获取漂浮的感觉。建筑形式为阿拉伯风格，充分运用了穹顶、花格栅和暖色调灰墁外墙等典型元素，并且将丰富的景观设计与建筑相结合，实现室内外环境的交融，体现了阿拉伯建筑的独到之处。

Be celebrated for its scenic golf course, the Lion Lake is chosen to be a world famous travel destination covering hotel, championship golf and high-end villa. The Sheraton Resort, which contains two guest room buildings, a conference center and a SPA, is the kernel of the whole complex. Raising the lobby to the third floor with a delicately designed hanging garden, it provides a great view of the lake. Guest rooms, bars and SPA are arrayed along the lakeside as well as the seafood restaurant and VIP rooms of SPA seating in the lake seem to float above the water. To capture the flavor of Arabic architecture, domes, patterned grilles, and stucco finish were incorporated into the design. The Arabic theme was carefully woven throughout the entire project by collaborating with the landscape and interior designers.

立面图

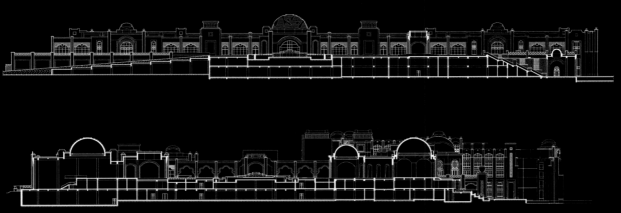

剖面图

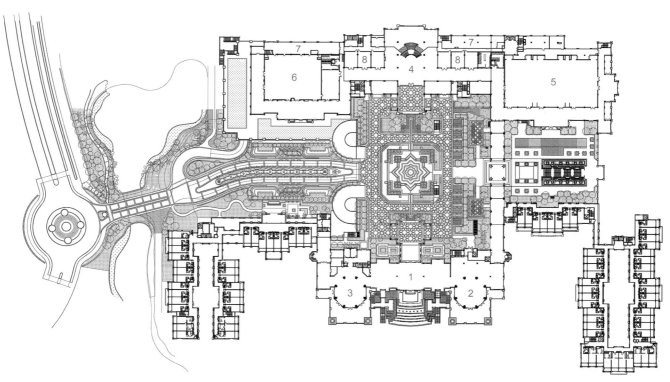

1. 酒店大堂 2. 接待厅 3. 休闲厅 4. 会议中心大堂 5. 会议大厅 6. 宴会厅 7. 厨房 8. 会议室

三层平面图

益田影人花园酒店　Cineaste Garden Hotel

地点 北京市怀柔区　/　**用地面积** 65 993m²　/　**建筑面积** 49 722m²　/　**高度** 80m　/　**设计时间** 2009年　/　**建成时间** 2012年

方案设计	李　凌　胡平淳
	马奕昆　徐伯君
设计主持	胡平淳　陈东红　杜爱梅
建　筑	徐伯君　马奕昆　王　静
结　构	贾文颖　彭永宏
给排水	王则慧
设　备	孙淑萍　郑　坤
电　气	李俊民
总　图	齐海娟
合作设计	澳大利亚五合国际

益田影人花园酒店及附属住宅是北京怀柔影视基地的配套项目，涵盖了五星级影视主题酒店、影视文化传媒中心、办公及影视文化配套商业等功能，可满足旅游观光、会议会展和商务交流的需要。其布局以五星级的影视主题酒店为中心，并为配合中长期居住客源的需求，相应设置了一定数量的办公空间。设计试图通过建筑语言来体现酒店的影视主题，将"红地毯"、"小金人"等影视界标志元素演绎成各种建筑构件，运用于整个建筑，使酒店具有鲜明的特色。 同时，设计也充分利用所在地区怡人的自然景观，以集中绿地及城市绿地营造出酒店的花园景观，烘托出项目的生态宗旨。

As an ancillary development of the movie & televsion base in Huairou, Beijing, the project includes a hotel, a media center, office and other facilities for film-television industry that could contains the activities of travel, conference, exhibition and business communications. The five-star theme hotel, which is the focal point of the complex, also obtains a few office spaces to meet the needs of long-term guess. A lots of symbols of film industry, such as red carpet and filmstrip, are transformed into various architecrtual components and ornaments. They actually make the hotel to be a theme park of film. In the same time, the design also pay attension to draw the view of local natural environment into the hotel.

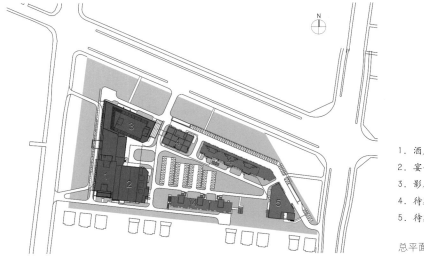

1．酒店
2．宴会厅
3．影人俱乐部
4．待建办公楼
5．待建商业

总平面图

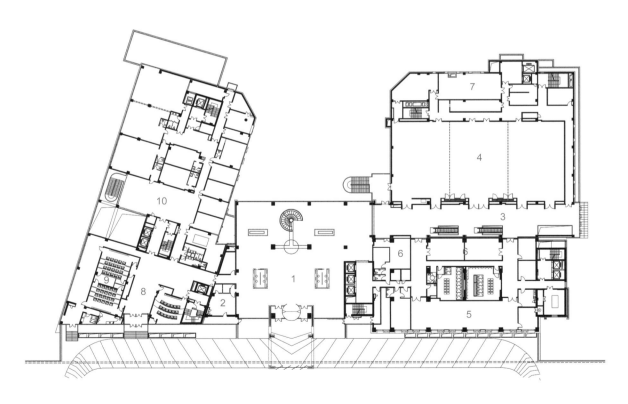

1. 酒店大堂 2. 服务台 3. 宴会区前厅 4. 宴会区 5. 小宴会厅 6. 休息厅 7. 厨房 8. 影人俱乐部前厅 9. 审片室 10. KTV区

二层平面图

中华人民共和国驻南非大使馆
The Embassy of the People's Republic of China in the Republic of South Africa

地点 南非比勒陀利亚市 / **用地面积** 24 701m² / **建筑面积** 13 595m² / **高度** 14m / **设计时间** 2004年 / **建成时间** 2011年

方案设计 崔愷 康凯 喻弢 肖晓丽
设计主持 崔愷 单立欣
建　　筑 康凯 喻弢 郑萌
结　　构 王文宇
给 排 水 王耀堂
设　　备 徐稳龙
电　　气 陈红 都治强
总　　图 白红卫
室　　内 张晔 顾建英 刘烨
景　　观 赵文斌

中国驻南非大使馆位于南非行政首都比勒陀利亚的近郊，通过逐渐收购相邻物业形成的较为宽敞的用地内。设计思路契合外交事务的风格，既要在异域体现中国气质，又要尊重当地文化，与邻里和谐相处。容纳办公、接待、签证功能的办公主楼布置在用地南侧，靠近主要道路，独立的签证入口面向东侧道路，外交官员的宿舍则位于用地西北部。与主入口连接的空间为对外接待部分，由南至北形成入口前庭、迎宾厅、休息平台、水院的空间序列，以圆形的月亮门作为收束，指向由舒缓的草坡组成的北侧花园。轴线两侧，通过建筑围合出一方方各具主题的庭院，并保留了基地原有的大树，在体现中国传统的同时留存了场地记忆。

The site of the embassy is a result after a series of acquisition for the nearby houses. In such a small-scale residential community, the building for diplomatic affairs should be modest to its neighbors and present Chinese atmosphere as well. The building encompassing offices, reception and visa affairs is seat on the south side of the site, while the dormitory of diplomatic staffs is on the northwest. Since there is an independent entry for visa, the main entrance is a ceremonial entrance which starts a spatial axis from the vestibule to the guest hall and a water courtyard and ended by a round "moon gate". Several courtyards on both sides of the axis present the traditional Chinese layout and contain most of the original trees to keep the memory of the site.

对梁架结构系统的强化，是对中国传统木构、英式维多利亚风格和南非土著建筑共有的木屋架传统的综合体现，表达了多种文化的融合。

The highlighted frames of roof structure show respect to the common factors of traditional Chinese, British and South Africa's wooden-structures.

建筑设计上刻意使以3600mm为模数的框架结构与墙体和屋盖系统相对独立，菱形屋架与双层顶构成的标志性元素在两端做出切角，而使混凝土屋架外露。金属遮阳格栅、中国特色的灰色面砖以及当地普遍采用的黑色石板瓦，兼顾了中国文化内蕴和对当地气候的回应。建筑框架的菱形图案，延伸到格栅、门扇、围墙和室内装饰中，形成内外一致的建筑语汇。

The 3600mm modular structure system is the dominant characteristic for the building. The corner cut roofs with rhombus concrete frames are relatively independent from the envelope system and the roofs. Both Chinese characteristics and local climates are responded by the metal sun-shade lattices, grey bricks and local black slates. The rhombus pattern is also extended into the design of exterior doors, enclosure walls and interior decoration that produce an integrated vocabulary.

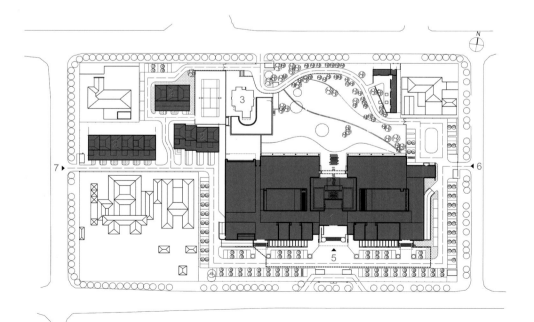

1. 办公主楼
2. 官员宿舍
3. 保留建筑
4. 辅助用房
5. 主入口
6. 签证入口
7. 员工入口

总平面图

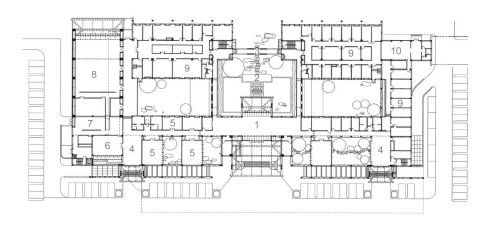

首层平面图

1. 迎宾厅
2. 水院
3. 月亮门
4. 侧厅
5. 会客室
6. 小宴会厅
7. 厨房
8. 多功能厅
9. 办公
10. 签证厅
11. 连廊
12. 办公中庭

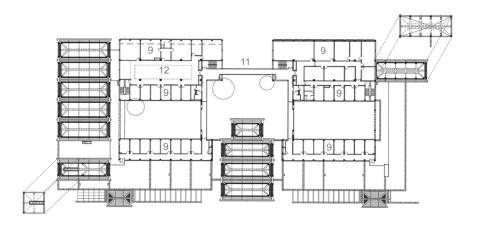

二层平面图

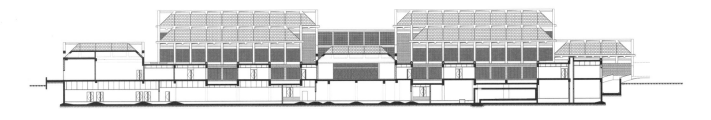

剖面图

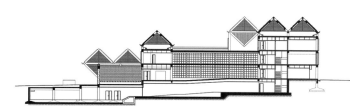

剖面图

中国驻开普敦总领事馆 The Consulate General of the People's Republic of China in Cape Town

地点 南非开普敦市 / **用地面积** 8 659m² / **建筑面积** 4 108m² / **高度** 12m / **设计时间** 2004年 / **建成时间** 2011年

方案设计 崔愷 喻弢
设计主持 崔愷 单立欣
建　　筑 喻弢
结　　构 王文宇
给 排 水 王耀堂
设　　备 宋玫
电　　气 贾金花
总　　图 白红卫
室　　内 张晔 刘烨 顾建英

中国驻南非开普敦总领事馆位于开普敦著名景点——桌山东南麓山脚，周边是茂密的山林和别墅庄园。用地为西北高、东南低的坡地，高差达13m，由在当地分别买入的三座院落组合而成。安检、签证、接待、办公、住宅等功能则分作五六个大小不一的体量，化整为零地布置在坡地上，达到融入环境的效果。

Surrounded by forest and country villas, the Consulate General is located on the foot of Table Mountain, the famous scenic spot of Cape Town. On the slope with a 13m-height-diffenece, the functions of the consulate such as security check, visa processing, reception, office and dormitory are contained in several small volumes. The massing strategy breaks the huge volume into parts and integrated to the environment.

三座院落之一为当地著名的历史建筑——名园，领馆方作为领事官邸保留原貌使用

One of the original buildings, Celebrated Garden, is a famous historical heritage of Cape Town and is conserved as the consulate's residence.

65

领事馆主入口和签证区与城市道路相连，经主入口安检区进入，顺地势而下，可由礼仪入口进入门厅。这里可与宴会厅连通，用以举办大型活动。两处厅堂面向入口草坪一侧均为落地窗，视野开阔，西北可望桌山山顶。地势较低处的内部办公区和馆员宿舍则背向外区，围合出内向的庭院，并设置了泳池。建筑造型兼顾中国文化内涵和南非当地特色，强调了"墙"和"四坡屋顶"两种元素：中式的墙用于阻隔视线和划分院落；西式的四坡顶则是当地常见的屋面形式。二者的搭配为建筑带来舒展的横向线条和丰富的层次感，营造出一种文儒书香、赏心悦目的宅院气息。

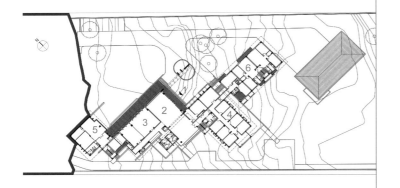

1. 主入口
2. 门厅兼多功能厅
3. 会客厅
4. 办公室
5. 签证办公
6. 馆员宿舍

入口大厅层平面图

7. 车行入口
8. 传达室
9. 签证厅
10. 后勤入口
11. 屋顶花园
12. 职工餐厅
13. 地下车库

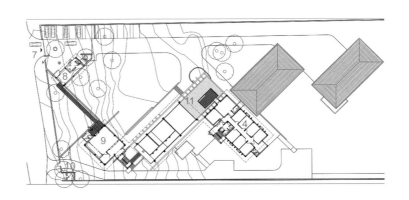

签证厅层平面图

The main entrance and the visa hall are seated nearby the city road. After a sloped approach passage, there is the reception area with a large lawn in front of it to provide a view of the Table Mountain and hold open-air ceremonies. The office part and dormitories enclose another court with swimming pool for staff uses. Considering "wall" as the element to block view and divide courts and "hipped roof" as a local architectural element, the building design combines them in an elegant way to present both Chinese and South Africa's features. The horizontal extending flat roofs and the layering effect give the building a welcoming and recognizable atmosphere.

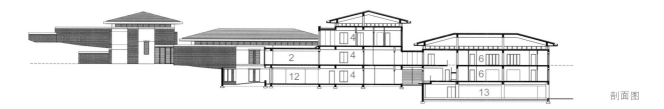

剖面图

新城大厦二期　Xincheng Building, Phase II

地点 江苏省南京市 / 用地面积 41 404m² / 建筑面积 177 387m² / 高度 89m / 设计时间 2009年 / 建成时间 2013年

设计机构　深圳华森建筑与工程设计顾问有限公司

新城大厦位于南京近年新城建设的重点区域，其位置与南京著名古迹"四方城"恰好对称处于南京市中心的东西两侧。为了强调这条连接历史和现代的空间轴线和视觉走廊，大厦被构思为一个"新四方"——以四个巨大的"城门"贯通城市轴线，以期成为新城区的地标建筑。工程分为两期建设：一期位于用地东侧，由两栋136m高的塔楼和裙房构成；二期为其主楼，两层高的U形裙房环绕主楼形成10m高的台地，对称布置的四栋80m高的办公主楼在顶部连通，构成"新四方"的主体形象。立面及平面设计均体现出严格而统一的模数体系，具有强烈的逻辑性及现代特征。

The location of Xincheng Building and the "Square City" goes through the Nanjing city on the direction of east and west amazingly. The "New Square" incorporates four giant "city gate", visualizing the axis in a dramatic way and becomes a landmark of the new district. The whole project is divided into two phases: the first phase locates in the east of the plot, the second phase locates in the middle, enclosed by two U-shaped podium buildings, which form an integrated landscaped terrace. The four 80m high towers are connected on the top, which composes the main image of the project. Strict modularized design of the façades and plans gives the building distinctive logical and modern characteristics.

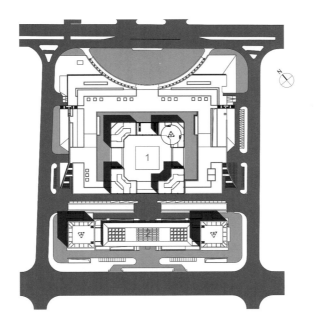

1. 二期建筑
2. 一期建筑

总平面图

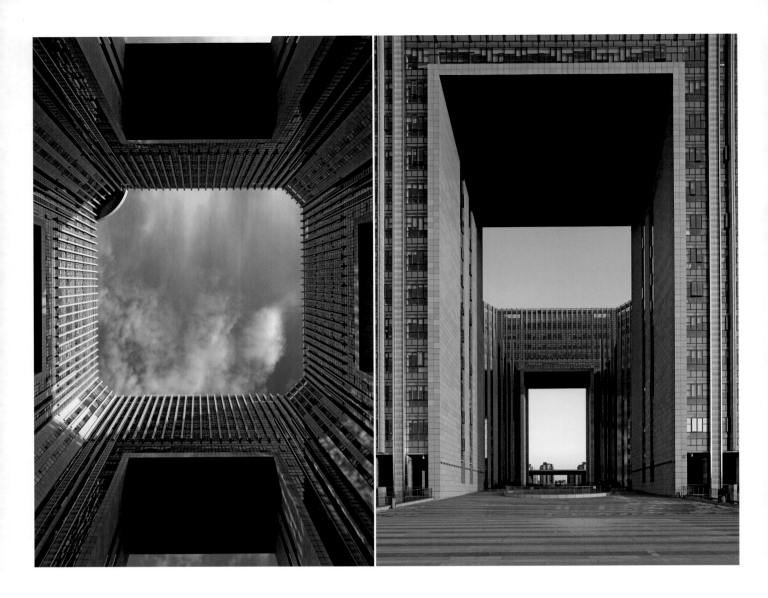

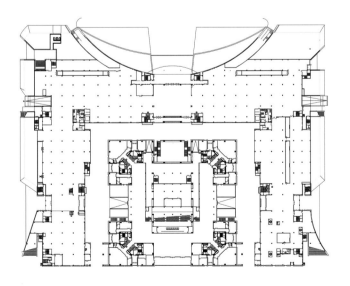

首层平面图

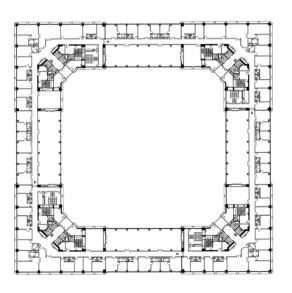

十五层平面图

南昌联发广场 Nanchang Lianfa Plaza

地点 江西省南昌市 / **用地面积** 31 600m² / **建筑面积** 220 000m² / **高度** 195m / **设计时间** 2006年 / **建成时间** 2012年

设计机构	中旭建筑设计有限责任公司
方案设计	曾筠 马文华
设计主持	曾筠 马文华
建　　筑	康红梅
结　　构	喻远鹏
给 排 水	孙路军
设　　备	郭晓楠
电　　气	姜晓先

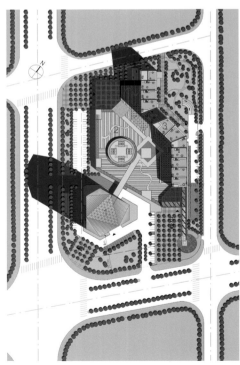

1. 办公塔楼 2. 商业 3. 公寓

总平面图

联发广场是一座集办公、商业及公寓功能为一体的城市超高层综合体，位于南昌市红谷滩区，与历史名胜滕王阁对江相望。尽管广场没有直接毗邻江岸，但通过将眺望江景的可能性最大化设计，扩大了观赏两岸人文和自然景观的视野。综合体内的商业布局，采取了传统的集中式商场与沿室内商业街店铺相结合的模式，公寓则以满足市场主流需求的小套型为主。对于作为地标形象出现的办公塔楼，在设计上以六边形的体量获得独特性，通过体形上适度的切削，获得如钻石形体般的意向，同时也契合了联发集团的企业铭训"钻石精神"。

A super-high urban complex incorporating office, retail store and apartment, Lianfa Plaza gestures to Ganjiang River beyond, and provides views of the famous historical building – Tengwang Pavilion. The commercial layout of the complex integrates the traditional department store and indoor commercial street. To satisfy the market demands, most units of the apartments are small-scale units. The hexagonal plan of the tower enhances the uniqueness of this landmark building. After moderate cutting of the top, the tower receives a crystalline feature, which also implies the motto of Lianfa Group, the "Diamond Spirit".

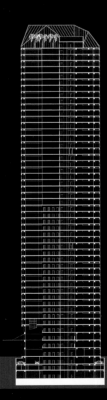

剖面图

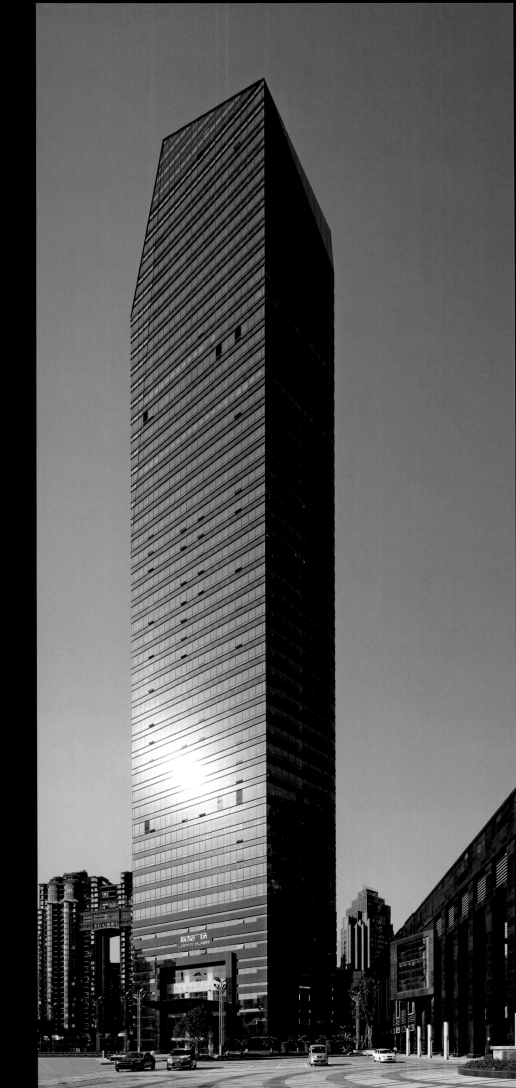

江西艺术中心　Jiangxi Arts Center

地点 江西省南昌市 / 用地面积 78 200m² / 建筑面积 63 618m² / 高度 40m / 设计时间 2006年 / 建成时间 2011年 / 座位数 大剧院 1500座

方案设计	张　祺　刘明军
	张　蓁　吴吉明
设计主持	张　祺　刘明军
建　筑	张　蓁　倪　斗　辛江莲
	胡　莹　吴吉明
结　构	王　载　刘建涛
	鲁　昂　王文宇
给 排 水	吴连荣　付永彬
设　备	汪春华　张　斌　梁　琳
电　气	马霄鹏　何　静
总　图	高　治
景　观	张　剑

江西的瓷器自古驰名中外。艺术中心的造型因而取自宋代瓷器经典的莲花纹样，以每一组花瓣对应着一种建筑功能——大剧院、音乐厅、排练厅和美术馆。基地东侧毗邻青山湖文化广场和风景秀丽的湖泊，艺术中心的入口广场因而与原有广场充分结合，形成形态完整的城市公共空间。建筑围绕广场布置，以大剧院和音乐厅为主体，美术馆设于用地北侧。组群起伏有致，各功能体之间的公共空间可分可合，便于分别经营，也提供了未来扩展的可能性。除了深远的寓意，建筑的莲花造型更形成了一种动势，层层递进的弧形墙面产生复杂的透视效果，以曲面增强建筑的情感表现力，也获得了室内外一致的丰富空间效果。多维非线性的石材外墙，采用60cm高的上下叠状的挂法，自然形成空间上的弧形，表面剔槽的方式加之小部分磨光，营造出细腻的反差效果。

As an art center of Jiangxi, the cradle of "china", the complex is designed with a reference to the classic lotus pattern of traditional porcelain. Each petal of the "lotus" houses a certain function: theater, concert hall, rehearse hall and arts gallery. Its urban design objectives included enlarging the Qingshan Lake Plaza it bordering by giving back an entrance square and enriching the urban space by its impressive curve feature. Enclosing the plaza, the ellipse "petals" provide a series public spaces between them, which could be divided or integrate to maximize the flexibility and future extension. A series of curvilinear walls not only create layered space effects, but also draw the dynamic sense into interior. The massive walls is faced with small scale plane granite plates, which collaborate together to form the three-dimensional spatial curved surfaces and show delicate effects by the contrast of sanding texture and polished texture of them.

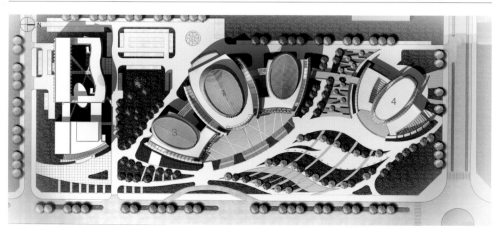

1. 大剧院　2. 音乐厅　3. 排练厅　4. 美术馆　　　　　总平面图

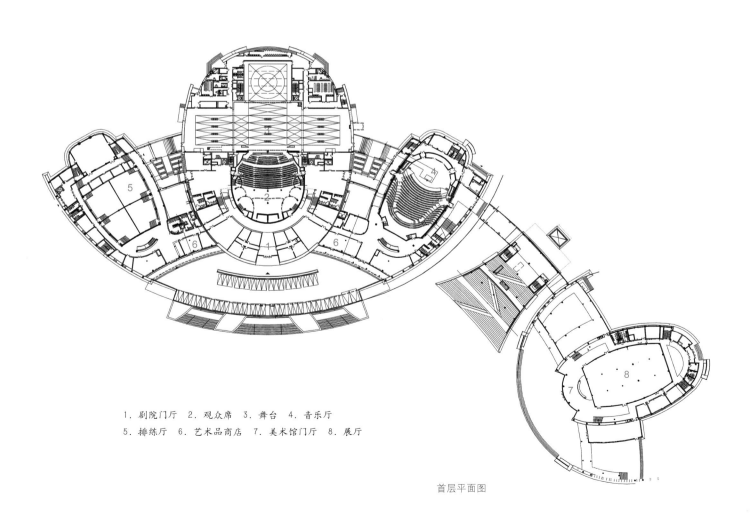

1. 剧院门厅　2. 观众席　3. 舞台　4. 音乐厅
5. 排练厅　6. 艺术品商店　7. 美术馆门厅　8. 展厅

首层平面图

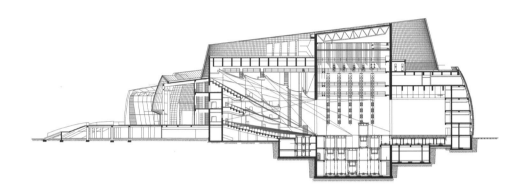

剧院剖面图

昆山市文化艺术中心　Kunshan Cultural Arts Center

地点 江苏省昆山市 / 用地面积 112 000m² / 建筑面积 72 410m² / 高度 31m / 设计时间 2009年 / 建成时间 2012年 / 座位数 大剧院 1400座

方案设计	崔愷 李斌
设计主持	崔愷 何咏梅
建　筑	李斌 张玉明 颜朝昱
结　构	朱炳寅 王奇
给排水	高峰 侯远见
设　备	金跃
电　气	陈琪 王烈
总　图	白红卫
室　内	张晔 纪岩 饶劢

水是昆山的灵魂，有三分之一的市域面积为水面覆盖。建筑坐落在以城市森林公园为核心的昆山西部副中心内，因此很自然地以曲线型的总体布局融入贯穿基地的水系形态。建筑体量由南侧向北面的森林公园逐渐舒展开来，滨水一侧设置了层次丰富的室外平台、休息厅、廊桥、散步道和广场，成为市民休闲娱乐的开放性场所。最能代表昆山文化的昆曲和并蒂莲成为建筑的母题，平面形态由昆剧表演的甩袖形象衍生而成，从一个中心逐渐旋转发散出来，其形态恰似盛开的莲花。

Located in the sub-center of Kunshan, a city well known for its large water area, the project develops its layout along the river in the site. A series of outdoor facilities facing the water surface, such as platform, resting area, bridge, walkway and plaza are defined as the urban space for citizens. Rotating and waving, the curvilinear geometries form two volumes that look like twin lotus flower, a symbolic plant of Chinese gardening. At the same time, the dynamic waves also intimate the movement of long sleeves in Kunqu Opera, which originates from Kunshan.

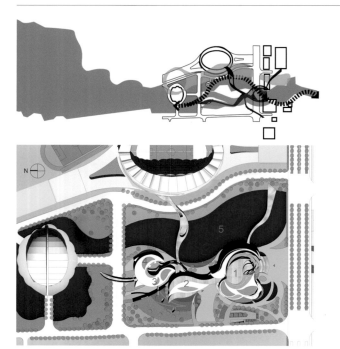

建筑形体来自对所处绿带水体和周边建筑的呼应。

The mass strategy derives from the surrounding green belt and water surface.

1. 大剧院
2. 影视中心
3. 文化活动中心
4. 展览中心
5. 景观水系

总平面图

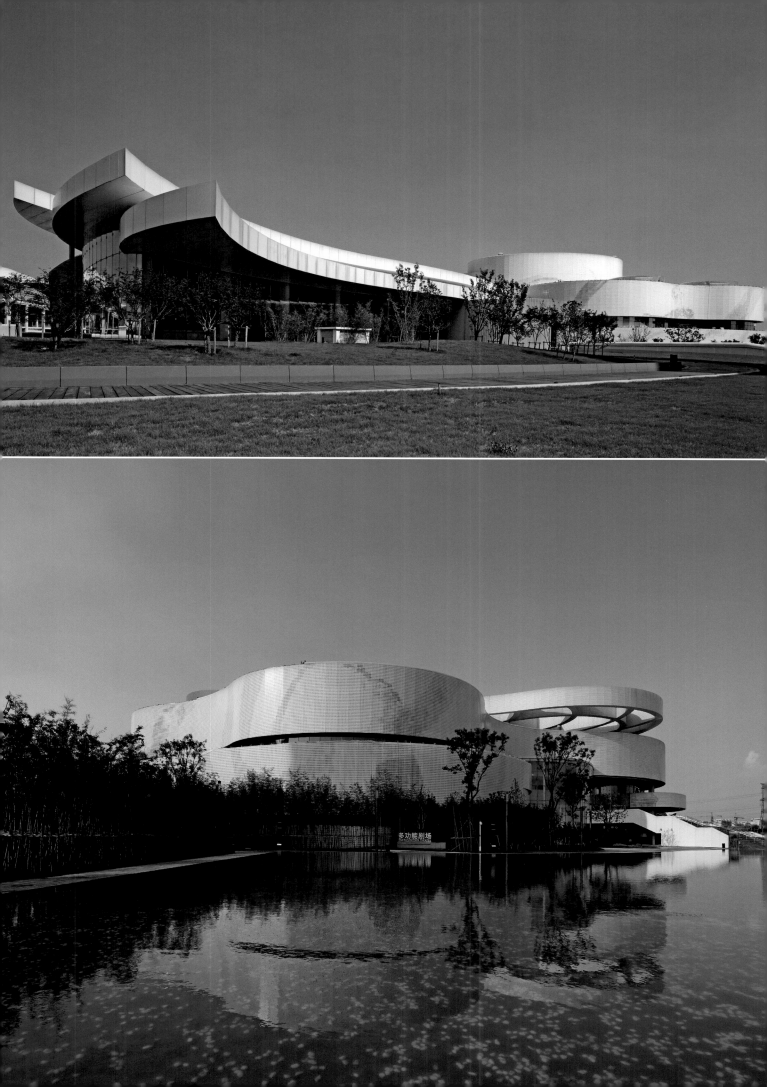

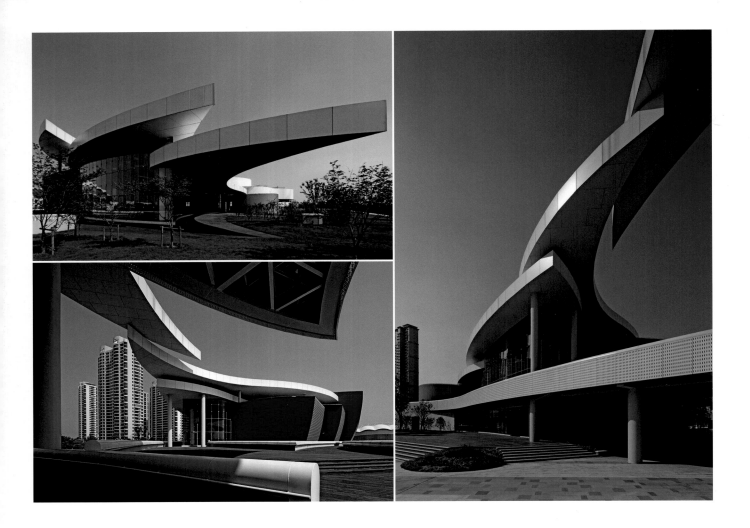

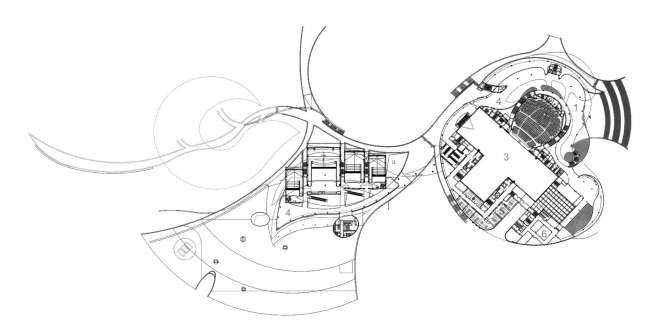

1. 门厅　2. 观众席　3. 舞台　4. 休息厅　5. 化妆间　6. 办公　7. 电影厅　　　　二层平面图

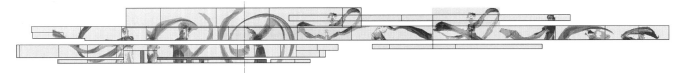

幕墙立面展开图

流线型的平台和楼板相互交错，凹凸有致的形态反映出江南水乡的灵动感。建筑外部覆盖白色调的穿孔金属板，通过孔洞的疏密变化调整透光率，在内侧形成通风的墙体，既提高了室内舒适度，其独特的图案变化也产生一种玲珑通透的肌理。

The interweaving and connection of horizontal floors also emphasize the ethereal atmosphere of traditional south China. Composed by perforated metal panels, the exterior surface wrapping the whole building presents a unique and translucent graphic pattern, while also creates a ventilation layer between the concrete walls and itself to improve the thermal comfort of interior.

德州大剧院　Dezhou Grand Theater

地点 山东省德州市 / 用地面积 59 740m² / 建筑面积 38 105m² / 高度 24m / 设计时间 2009年 / 建成时间 2013年 / 座位数 歌剧院 1500座

方案设计	李燕云　赵丽虹　辛江莲 罗　洋　赵梓藤
设计主持	李燕云　赵丽虹
建　　筑	王　斌　关　帅
结　　构	孙海林　罗敏杰
给 排 水	马　明
设　　备	刘筱屏　李　嘉
电　　气	熊小俊　任亚武
总　　图	白红卫
景　　观	李　力

德州大剧院由一个大型歌剧院和一个多功能小剧场组成。总体布局采用与周边环境相契合的曲线形态，两个椭圆形体量由一个贯通南北的共享大厅相连，顺应场地现状，向南形成环抱之势。建筑弧形外表皮优美流畅，结合内部复杂的使用功能，形成了富有韵律的音符造型，灵动而具有节奏感。建筑外围的东西两侧共设置了40榀渐变的穿孔金属帆片，有序地沿弧线展开，宛如两组徐徐拉开的舞台大幕，将建筑的艺术特质展露无遗。

Composed by a grand theater and a little multi-functional theater, the complex also employs a mass strategy with two ellipse volumes to embrace an entrance square, while a concourse connects both of the ellipses. The curvilinear skins outside the volumes provide rhythm and dynamic effects. The 40 "sails" penetrated with gradient changing patterns arrayed perpendicularly to the ellipse volumes. Silhouettes of those metal units imply the huge curtains of stage to point out the function of the artistic facility.

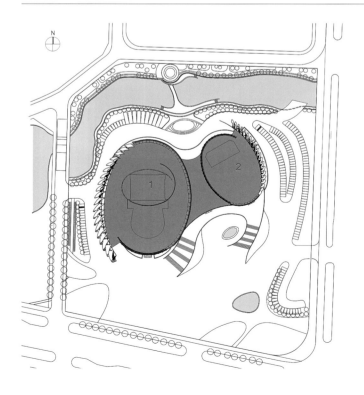

1. 歌剧院
2. 多功能剧院

总平面图

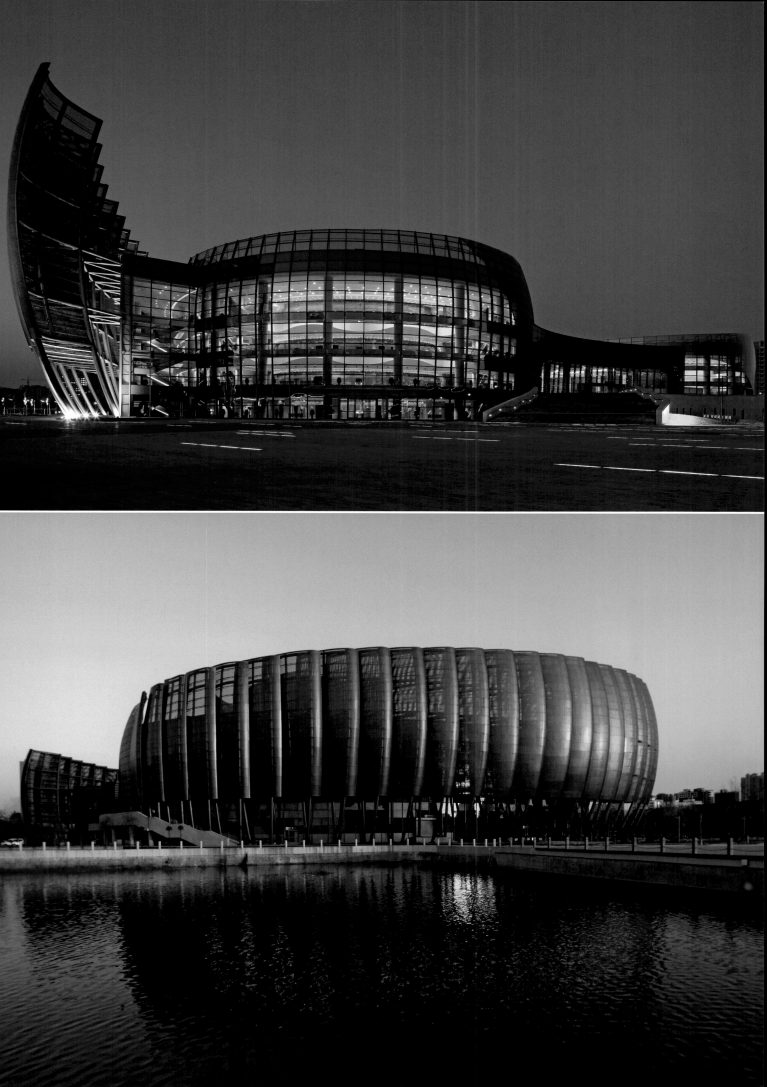

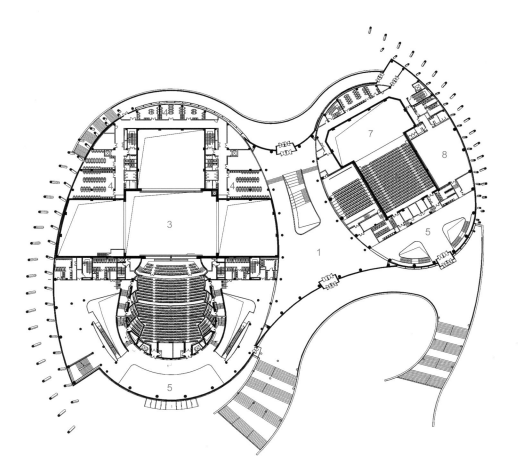

1. 城市共享大厅
2. 歌剧院观众席
3. 歌剧院舞台上空
4. 化妆间
5. 休息厅
6. 多功能剧场观众席
7. 多功能剧场舞台上空
8. 排练厅

二层总平面图

青海艺术中心 Qinghai Arts Center

地点 青海省西宁市 / **用地面积** 36 017m² / **建筑面积** 30 506m² / **高度** 35m / **设计时间** 2007年 / **建成时间** 2012年 / **座位数** 剧场 1200座

方案设计	张祺 刘明军 宋菲 辛江莲 吴吉明
设计主持	张祺 刘明军
建　筑	宋菲 辛江莲
结　构	尤天直
给 排 水	夏树威 刘海
设　备	蔡玲 李冬冬
电　气	李维时 郭利群
总　图	连荔

青海艺术中心主要由一个可容纳1200人的中型剧场和一个800座的音乐厅，以及若干相关服务设施组成。两个椭圆形体量分置于建筑基座的南北两侧，观众由入口台阶登上二层平台，经由两者之间的城市共享大厅进入剧场及音乐厅。东西向通透的大厅为观众提供了连贯海湖景观带的开敞视线，周边设置的餐饮、服务功能则使之成为一处市民休憩、交往的场所，同时可进行展览、艺术讲座等活动。艺术中心的设计以高原雪山为造型意象，两个主要椭圆形体量的外墙向内倾斜，呈盘旋上升的螺旋状，加之有节奏设置的开窗和通长的竖向格栅，增加了虚实变化，使整个造型呈现出音乐般富有韵律的灵动感。

The arts center is composed of a 1200-seat middle-sized theater, an 800-seat concert hall and other relative facilities. On a massive podium, two elliptical volumes anchor the complex at north and south ends and provide an entrance enclosed by cable-supported glass walls between them. Bending inward, the sweeping curtain wall invites citizens to the concourse and keeps the views to the urban landscape belt from any interruptions. The catering and service facilities in the concourse make itself an open space for communication and rest. A lot of artistic activities also occur here frequently. The spirals of the massive walls, which wrap the ellipse volumes and the podium, indicate the soaring snow mountains of Qinghai.

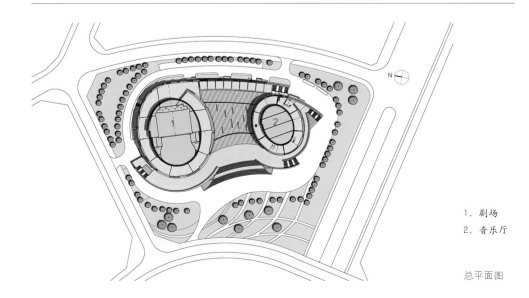

1. 剧场
2. 音乐厅

总平面图

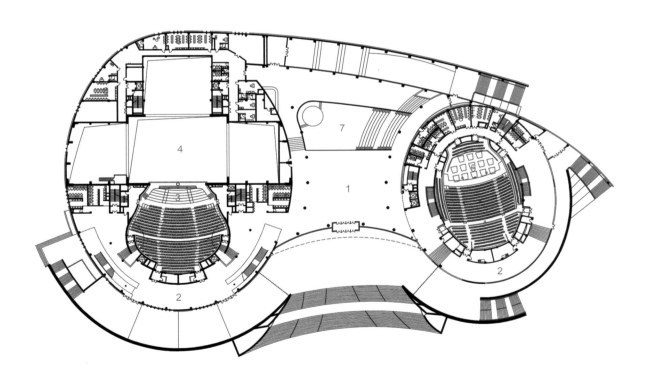

1. 城市大厅 2. 休息厅 3. 影院观众席 4. 舞台 5. 音乐厅 6. 化妆间 7. 下沉演艺厅

二层平面图

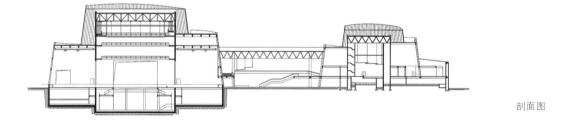

剖面图

大沽口炮台遗址博物馆　Dagukou Fort Ruins Museum

地点 天津市塘沽区 / **用地面积** 54 718m² / **建筑面积** 3 710m² / **高度** 11m / **设计时间** 2005年 / **建成时间** 2011年

方案设计	崔　愷　文　兵　白　晶
设计主持	崔　愷　张　波
建　　筑	白　晶
结　　构	张淮湧
给 排 水	申　静
设　　备	刘筱屏
电　　气	李战赠
总　　图	郝雯雯

大沽口位于天津海河入海口，设防于明嘉靖年间，是京畿的海上门户。其炮台建于清嘉庆年间，曾在第二次鸦片战争及八国联军入侵时曾多次抵御外国侵略者。博物馆的设计突出了炮台这一遗址保护的本体，建筑作为环境中的从属，其本身具有强烈冲击感和沧桑感的形式，也成为展陈的一部分，唤起人们对那段苦难历史的回忆。水平展开的建筑如同从大地中生长出来，突出而非削弱炮台的主体地位。放射状裂开的形式看似高低不同、相互交织，实际来源于炮台等标志性场所与博物馆的视线联系，使炮台的景观成为展陈内容的有机组成部分。

Lying 45 kilometers southeast of Tianjin City, the Dagukou Fort was built in 1817 to protect the capital, Beijing. As a heroic symbol of China's fight against foreign invasion, especially the Second Opium War and the Invasion of the Eight-Power Allied Forces, the emplacement requests an unassumbling but impressive building to contain its memory of the invaded history. To emphasis the forts' major status in the site, the museum did not obtain a complete image. At first sight, the horizontally developed mass emerges as gaps of the earth, while the random scattered layout is actually a result to form visual relations between the forts and interior spaces. By this means, the scene of fort becomes an organic part of the exhibition.

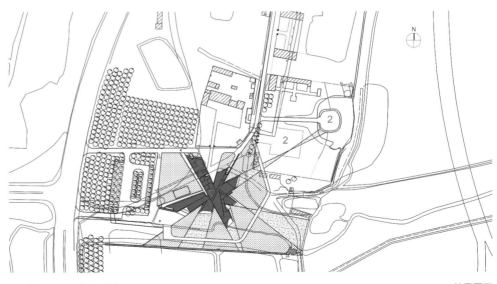

1. 博物馆　2. 炮台遗址　　　　　　　　　总平面图

展览流线结合遗址整体参观需要,围绕炮台组织,以突出其地位。建筑外墙上色泽凝重的锈钢板,恰如一门门现场出土的锈迹斑斑的铁炮,诉说着一段中华民族屈辱而又不断抗争的历史。

The whole visiting circulation is organized around the forts, which is certainly the centerpiece of the site. Wrapped with weathering steel plates, the exterior of the museum is an extension of the rusted canons, conveying a vivid feel for the cruel battle that took place in this area.

1. 门厅
2. 服务台
3. 展厅
4. 放映厅
5. 库房
6. 办公

平面图

蓬莱古船博物馆 Penglai Ancient Ship Museum

地点 山东省蓬莱市 / 用地面积 8 503m² / 建筑面积 7 276m² / 高度 6m / 设计时间 2006年 / 建成时间 2012年

方案设计	崔愷 康凯 傅晓铭
设计主持	崔愷 张男
建 筑	赵晓刚 张汝冰
结 构	贾卫平 赵莉华
给排水	高峰
设 备	刘玉春 何海亮
电 气	甄毅
总 图	白红卫

古船博物馆位于蓬莱水城保护区范围内,在遵循文物保护基本原则的前提下,设计构思希望再现古军港帆樯林立、战舰森森的壮观景象,以及船只修造的场面,并将游客体验式参观和建筑功能及流线组织结合起来。沉船遗址展厅位于船骸原址之上,其主体埋入地下,地面仅保留绿坡和屋顶的复原古船台展示场。管理研究用房是相对独立的地面仿古院落,与水城内的复原古建筑群相呼应,地下与博物馆的展区相通,从而保持了水城风貌的历史真实性和完整性。

Situated in the water town conservation zone of Penglai, the site museum is started to recur the spectacular of ancient military harbor with sails and masts. The visiting circulation provides a series of experiences for the shipbuilding and military and connects other service facilities. Covered by the exhibition platform of rebuilt ancient ships and grass slopes, the site hall is embedded into the earth modestly. A set of traditional style courtyards is set relatively independent in the northwest corner of the site to show respect to the whole ancient building complex of the water town, while its underground level is connected to the exhibition hall.

总平面图　1. 博物馆　2. 蓬莱阁　3. 总镇俯　4. 三官庙　5. 小海　6. 黄海

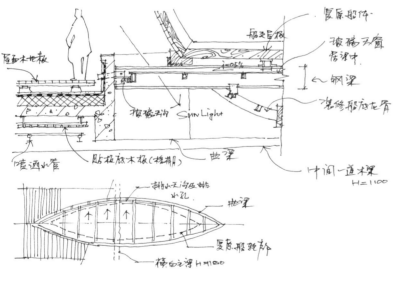

船台构造示意草图

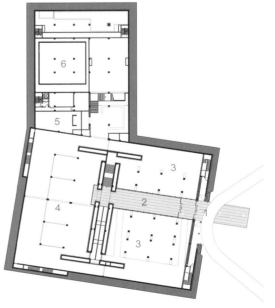

1. 游客入口 2. 序厅 3. 复原遗址上空
4. 文物展厅上空 5. 报告厅上空 6. 文物库房上空

首层平面图

置身斜柱支撑的仿船坞空间，观众可以在入口序厅层与底层这两处不同高度，两次看到残船。建筑空间结合展陈设计，将残船文物、复原影像以及顶棚的水下船底意象并置展示，多层次地凸现古船主题。

The V-shaped structural supports of the site hall bring a sense of dockyard. By entering the hall and stepping down to the underground level, visitors encounter the ship ruins from different angles. The ruins, the restored image and the ship bottom image on the ceiling collaborate together to embody the ancient ship in multiple aspects.

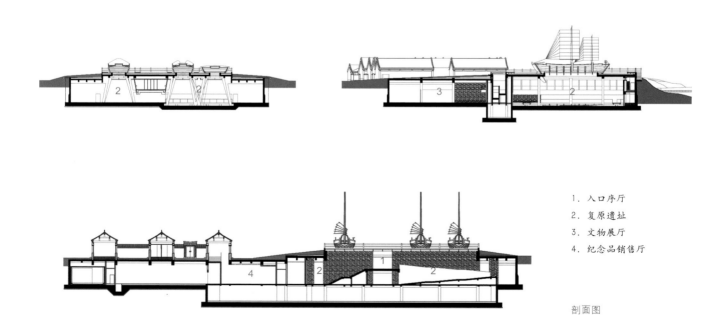

1. 入口序厅
2. 复原遗址
3. 文物展厅
4. 纪念品销售厅

剖面图

中国杭帮菜博物馆 Chinese Hangzhou Cuisine Museum

地点 浙江省杭州市 / **建筑面积** 12 470m² / **高度** 12m / **设计时间** 2010年 / **建成时间** 2012年

方案设计 崔愷 吴朝晖 周旭梁
设计主持 崔愷 周旭梁
合作设计 浙江大学建筑设计研究院

杭帮菜博物馆是集展示、体验、品尝"杭帮菜"功能于一体的主题性博物馆。作为地方饮食文化的载体，其设计用现代的材料体现了杭州"秀、雅"的神韵。建筑所在的江洋畈生态公园，原为西湖疏浚淤泥库区，经过将近十年的表层自然干化，已形成与周围山林不同的，以垂柳和湿生植物为主的次生湿地。为了削弱建筑对生态公园的压迫感，建筑体型随山势和地形蜿蜒转折、自然断开，划分成贵宾楼、餐饮区、博物馆经营区和固定展区等四个功能组团。

A place for exhibition, experience and taste, the Hangzhou Cuisine Museum is designed with modern construction mode and materials to recall the elegant quality of Hangzhou. The building is located in a valley used to stack the sludge from West Lake. After 10 years, the surface of sludge is dried and grows weeping willows and wetland plants. In such a wetland park with unique scenery, the building is developed along the foot of Qianwang foot and split into four parts: VIP house, dining hall, museum store and exhibition hall.

1. 博物馆固定展区 2. 博物馆经营区 3. 餐饮区 4. 贵宾楼 5. 公园湿地 6. 钱王山   总平面图

建筑体量的拆分,削弱对公园的压迫感,也保留了公园与钱王山之间的视觉通廊。连续折面坡屋顶的形式,进一步减小了建筑的尺度,形成统一而又富有韵味的形体和空间变化,并与自然山体轮廓相呼应。绿色植草屋顶也使建筑真正地融入环境之中。

The decreased mass minimizes the building's depression to the park as well as retains the visual channels from the wetland to Qianwang Mountain. Inspired by the undulation of nearby mountains, the continuous slope roofs diminish the building scale and gain a series of massing and spacial variation. At the same time, the planting roof is another method to make the building a part of the natural environment.

各功能组团之间以木栈道和休息木平台连接成整体,这些景观元素同时也是整个公园木栈道系统的组成部分,可供游客休息、观景之用,长长的屋顶挑檐的遮蔽,使得室内活动的空间能够延伸到室外水边和公园之中。

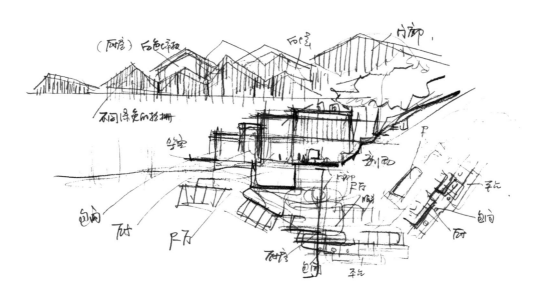

设计草图

Wooden trestles and platforms are not only elements to connect the functional parts, but also components of the whole wooden trestle system for the park. Shaded by the overhanging roofs, this area becomes a transition between interior and exterior.

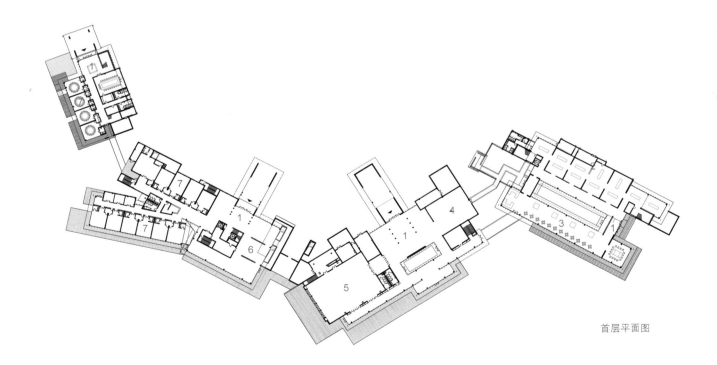

首层平面图

1. 门厅 2. 固定展区 3. 展览体验区 4. 售卖区 5. 多功能区 6. 餐厅 7. 包厢

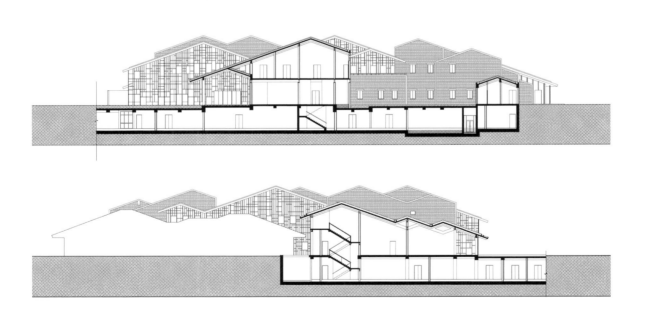

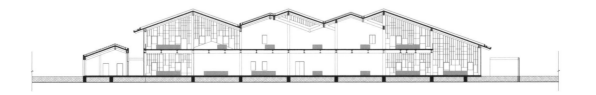

剖面图

濮阳市城乡规划展览馆　Puyang Planning Exhibition Hall

地点 河南省濮阳市　/　**建筑面积** 15 000m²　/　**高度** 20m　/　**设计时间** 2008年　/　**建成时间** 2012年

方案设计	郑世伟　吴燕雯　赵园生
设计主持	郑世伟
建　　筑	白　芳　李盈瑞
结　　构	朱炳寅　刘　巍
给 排 水	吴莲蓉
设　　备	魏文宇
电　　气	许冬梅
总　　图	郑爱龙

这座因中原油田而繁荣的城市，本身也是一座历史文化名城。设计尝试从老城区中选取一处典型的城市公共空间，将其舒适、利于交往的街道空间尺度，移植到城市规划展览馆中，让参观者通过这一空间片段，真切地感受到濮阳特有的城市状态和氛围。体形方正的建筑内部，几条"街道"指向实际上的城市门户和中心广场，以此暗示建筑和城市的内在联系，促使旅游者与城市居民汇聚于此，共同了解这座城市。除了展示城市建设成果，建筑内部还提供城市服务、展览、培训和辅助商业功能，成为一处开放的文化场所。

The city flourished as the center of an oil field, Puyang, is a historic city in itself. For the planning exhibition hall, to form a montage of the old down town will be an appropriate way to record the comfortable street spaces and lives, which are the most valuable merits of the city. In the pure cube, the directions of the diagonal streets point out the location of the city gate and urban plaza to indicate the connection between the building and the city. They also could be considered as clues for travelers to come and communicate with the citizens here. Besides the exhibition of city planning, there are a series of other functions, such as community service, exhibition, training and relative retail stores.

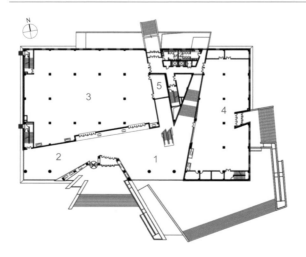

1. 门厅　2. 休息空间　3. 展览厅　4. 报建服务厅　5. 库房

首层平面图　　　　　　　　　　　　　　　　　　　　设计构思

宁波国际贸易展览中心2号馆 Ningbo International Trade & Exhibition Center No.2 Pavilion

地点 浙江省宁波市 / 用地面积 42 400m² / 建筑面积 201 148m² / 高度 48m / 设计时间 2008年 / 建成时间 2010年

方案设计　徐　磊　曹晓昕　白　晶
设计主持　徐　磊　安　澎
建　　筑　丁利群
结　　构　王春光
给 排 水　陈　宁
电　　气　王浩然
总　　图　连　荔

宁波国际贸易展览中心2号馆位于同为本院设计的1号馆南侧。交通设计是会展建筑的关键。展馆的大部分人行入口集中于东、南两侧，与景观区相连，便于人流疏散，减少人车流干扰；车行交通则集中于西、北两侧，以立体交通的方式解决了北侧办公和展览流线交叉的问题。立面设计使上下部分有所变化，以曲折的顶部形态削弱了建筑体量。建筑形体由外部遮阳金属构件统一，同时也产生了丰富的细节。4个通高的中庭空间，起到了促进空气对流、增加自然采光的效果；而沿建筑长向的室外通廊则减少了建筑进深，增强自然通风效果，并有利于遮阳。

Just seated to the south of the No.1 Exhibition Pavilion, the No.2 Pavilion is designed to emphasis the circulation system of the building. Most of the pedestrian entrances are located on the south and east sides near to a landscape, while the vehicle entrances are set on the north and west sides. An apparent gap between upper levels part and the lower levels on the façade demonstrates the functional division of section. Part of the upper levels is cut to reduce the huge volume of the building. As a dominating element of the façades, the grilling also gains delicate effects of light and shadow. In interior, four atria create stack effect and admit natural light. The outdoor corridors decrease the depth of plan to enhance natural ventilation.

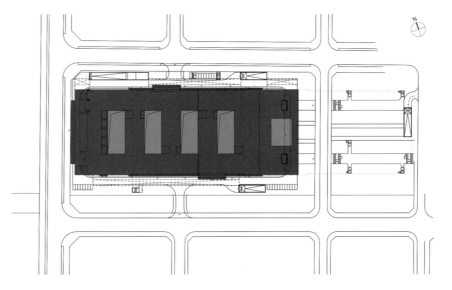

总平面图

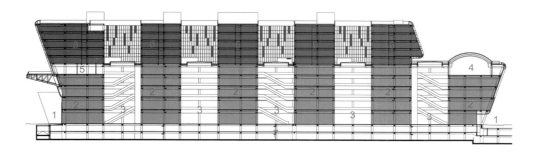

1. 人行入口 2. 展览区 3. 中庭 4. 会议厅 5. 餐饮区 6. 办公区 7. 停车卸货区

剖面图

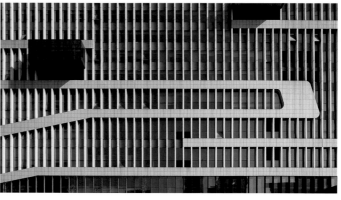

青海科技馆 Qinghai Science & Technology Museum

地点 青海省西宁市 / 用地面积 36 600m² / 建筑面积 33 179m² / 高度 33m / 设计时间 2007年 / 建成时间 2011年

方案设计　张　祺　史秋实　班　润
设计主持　张　祺　刘明军
建　　筑　张　蓁　史秋实
　　　　　杨鸿霞　班　润
结　　构　王春光　尤天直
给 排 水　吴连荣　朱跃云
设　　备　张　斌　马　豫
电　　气　李陆峰　郭利群
总　　图　连　荔

青海科技馆集中了科普教育、展示、交流培训等多种功能，位于西宁市海湖新区中心区内，与东侧的青海艺术中心呈环抱形之势。建筑的南北向均可远眺壮观的群山，为了体现高原特有的地域文化，整个建筑在体量上强调平衡、稳定之感。主要功能分为四个部分：中心大厅、展廊、展馆及辅助用房，根据其内在功能采用相应的建筑空间形态，形成了丰富的室内外空间效果。中心大厅的穹形空间、展馆的门式框架结构、展廊的采光顶，体现了结构的整体感和韵律感。以建筑形式产生出时空隧道、浩瀚天宇、科技长廊等意象，与其功能相呼应。

An education base for science popularization, exhibition and communication, the Qinghai Science & Technology Museum is lie on the west of the central city plaza of the new urban district, as a counterpart to the Qinghai Arts Center. Not far away from the grand mountains, it values the stable and distinct feature of the plateau terrain in its mass strategy. Four main functional areas, central hall, gallery, exhibition hall and service rooms are accommodated in different geometries, such as ellipsoid, curved tube and filleted box for lecture room, which are echoes to the functions and also offers space effects fitting for the cosmical and high-tech scenes.

总平面图

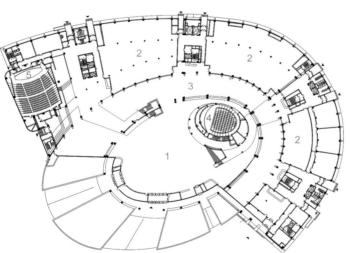

1. 中心大厅
2. 展厅
3. 展廊
4. 放映厅
5. 报告厅

首层平面图

鄂尔多斯机场新航站楼　New Terminal of Erdos Airport

地点 内蒙古鄂尔多斯市 / **用地面积** 150 000m² / **建筑面积** 100 277m² / **高度** 31m / **设计时间** 2008年 / **建成时间** 2012年

方案设计　中旭建筑设计有限责任公司
　　　　　　加拿大B+H建筑事务所
设计主持　曲雷　韩玉斌
建　　筑　孙净
结　　构　贾卫平　李忠盛　郭海山
给 排 水　王涤平
设　　备　魏文宇
电　　气　马宁
电　　讯　罗卫东
总　　图　齐海娟
室　　内　何勍
合作设计　中国民航机场建设集团公司

鄂尔多斯机场新航站楼以"草原雄鹰"的形象，生动地体现出这个快速崛起的能源城市的特点。其设计综合考虑了航空、陆地、轨道交通的对接，并按照功能流线清晰地划分为到港层、离港层和容纳商业餐饮设施的夹层。跨度达108m的圆形离港大厅如同草原上的大帐，使旅客一进入机场就感受到强烈的震撼。中部直径20m的天窗根据富勒球原理计算，使壳体结构与天窗外框完全统一，获得了通透的效果。穹顶下的油画长卷则展现了成吉思汗的一生。大厅左右两组540m长的曲面双翼，覆盖着开阔的候机大厅。建筑中大量暴露的结构构件，其尺寸、形态均经过细致推敲，V形柱头减小了结构跨度，室内棱形柱则因刻意隐藏了水平梁而显得修长、俊美。

The Eagle of the Grassland, a very metaphor for the new terminal of Erdos Airport, shows an iconic image for the rapid developing city. As a joint of air travel, roadway and metro, the terminal clarifies its function into three stories: the arriving level, the departing level and a mezzanine for retail and restaurants. With a column-free span of 108m, the concourse, which is an iconic image of the airport, gives a deep impression to passengers. After a study of Fuller Dome, the frame of the central skylight becomes a total combination of architecture and structure to maximum the visual openness. Under the ceiling is a long oil painting describing the life of Genghis Khan. Two wings with 540m-long curved canopies containing waiting halls flank the concourse. By articulated exposition, the structural components present the beauty of structure design.

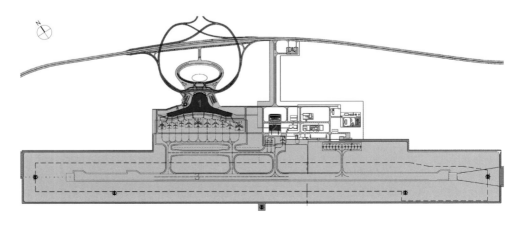

1. 新航站楼　2. 原有航站楼　3. 跑道　　　　总平面图

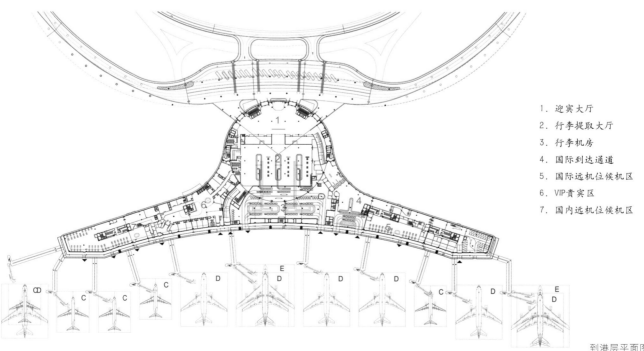

1. 迎宾大厅
2. 行李提取大厅
3. 行李机房
4. 国际到达通道
5. 国际远机位候机区
6. VIP贵宾区
7. 国内远机位候机区

到港层平面图

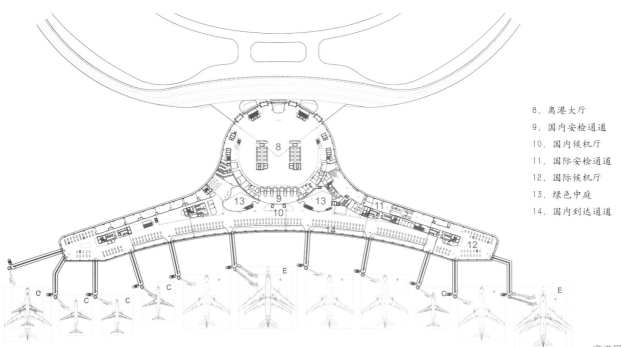

8. 离港大厅
9. 国内安检通道
10. 国内候机厅
11. 国际安检通道
12. 国际候机厅
13. 绿色中庭
14. 国内到达通道

离港层平面图

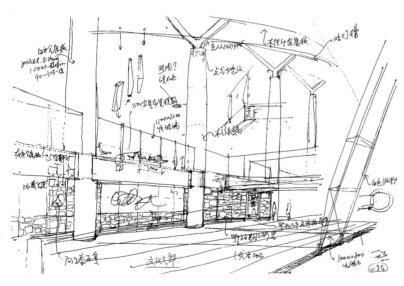

手绘草图——对室内装修的考虑

手绘草图——天窗

进出港必经之处的穹窿下，设置了一对被称为"绿肺"的中庭，不仅充分引入阳光和当地稀缺的绿色植物，也在其周边自然结合了旅客进出港通道、候机区、咖啡厅、儿童活动区等众多服务设施，使中庭成为各类功能活动的中枢。以天空和白云作为意向的迎宾厅体现了草原的意向，贵宾厅则以传统蒙古族纹样进行装饰。

A pair of top-lit atria, which contain lush interior landscapes enlivened by plants and sunlight, are added into the junction of the concourse and two wings. Since lots of functions for passenger activities, such as the waiting area, café, children playing space are organized around them to share the view and sensual atmosphere, the atria become the centerpiece of the terminal.

剖面图

大同机场新航站楼 Datong Airport New Terminal

地点 山西省大同市 / 用地面积 107 380m² / 建筑面积 10 854m² / 高度 32m / 设计时间 2010年 / 建成时间 2013年

方案设计	崔 愷 杨金鹏
	刘 德 闫小兵
设计主持	杨金鹏
建 筑	陈帅飞 刘 德 闫小兵
结 构	杨 苏
给 排 水	黎 松
设 备	刘玉春
电 气	都治强
电 讯	任亚武
总 图	高 治
室 内	张明杰
景 观	雷洪杰
合作设计	中国民航机场建设集团公司华北分公司

大同机场新航站楼采用前列式布局，体量规模并不大，但是采取小中见大的策略，将各个功能集成于斜向屋顶之下，获得了开敞宽阔的室内空间效果。出发大厅、候机大厅、行李提取大厅通过开放的中庭和屋顶桁架采光廊对话，陆侧和空侧隔而不断，空间连贯。建筑造型提取了山西大同传统建筑硬山屋脊的元素，采取大跨度非对称人字形桁架构成结构主体，倾斜的金属屋面材质和构造极富现代感，空间和气质古拙浑厚，结构形式与建筑造型，建筑功能与室内外空间相互对应形成一体化，形成了和通常的机场航站楼截然不同的形象。

Although it is a middle-sized terminal, the building obtains a sense of openness by concentrating most of the functional areas under the upper waiting halls. Inspired by the gabled roof of traditional Datong buildings, the terminal is dominated by an asymmetrical slopping truss and gains a large column-free interior space. The departing hall, waiting hall and baggage claim hall are arranged close to the spacious concourse lit by skylights beyond the ridge. Visual obstructions between airside and landside are also minimized to keep an orienting view. Faced with metal plates, the roofs still features an image far away from typical airport terminal appearance that echoes the past and developed from the modern architectural and structural manners.

大同华严寺大雄宝殿屋顶

航站楼被构思为陆侧和空侧之间的转换器,其形态体现出发的动势,同时也参考了大同当地壮丽奇特的山地景观,如同从大地中升起,成为天地间对话的节点。

The terminal is conceived to be a converter of airside and landside and an imagery of flight by its dynamic gesture to the airside. Emerging from the earth, it also could be look like a node of the conversation between sky and earth.

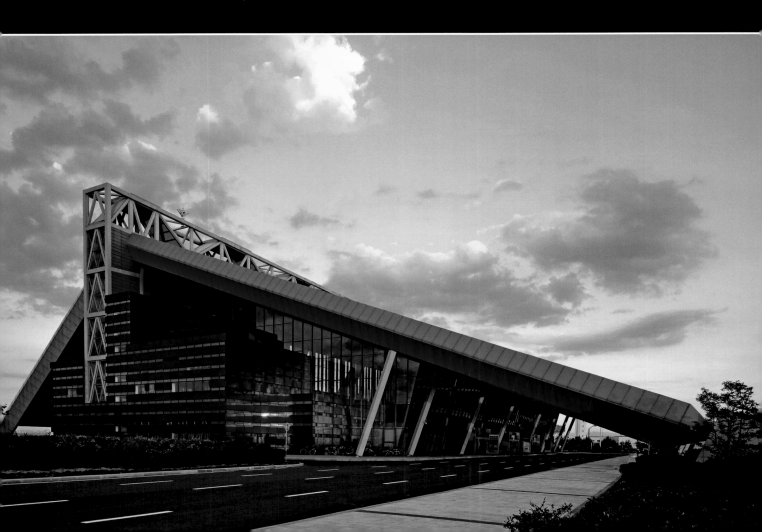

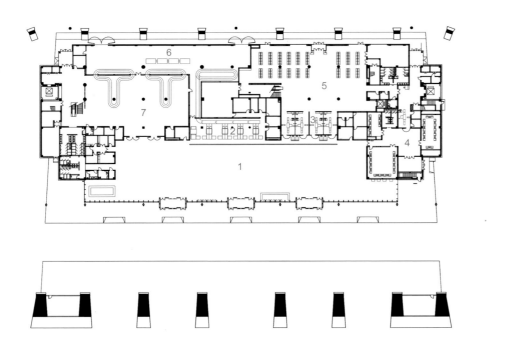

1. 综合迎送大厅
2. 值机柜台
3. 安检
4. 贵宾区
5. 远机位候机厅
6. 行李到达分检房
7. 行李提取大厅

首层平面图

剖面图

苏州火车站站房 The Building of Suzhou Railway Station

地点 江苏省苏州市 / **用地面积** 96 000m² / **建筑面积** 85 800m² / **高度** 31m / **设计时间** 2006年 / **建成时间** 2013年

方案设计	崔愷 王群 李维纳 狄明 贺小宇 涂欣 章春
设计主持	崔愷 王群 李维纳
建筑	王喆 龚坚 狄明 涂欣
结构	范重 张亚东 彭翼 赵长军
给排水	匡杰 苏兆征
设备	刘玉春 何海亮
电气	李俊民
总图	连荔 徐忠辉
室内	顾建英 郭晓明 李晨晨
景观	李存东 赵文斌 王洪涛
合作设计	中铁第四勘察设计院集团有限公司

苏州站位于老城北边，隔护城河与姑苏相望。现代化轨道交通系统的发展使老车站难以满足使用要求，需要一个大规模、综合性的交通枢纽取而代之。新站创作的核心问题是如何使庞大的空间体量与苏州细腻、幽雅的小尺度氛围相协调，从而坚持了"苏而新"的本土建筑原则。设计将菱形作为一个符号系统进行发展，从大跨度的站房空间桁架体系，到门窗檐口以及地面铺装的不断演绎，再配以白墙黛瓦的苏式淡彩，还有大大小小的苏式庭院和候车敞廊，使这个庞大的现代化车站能够与古城对话，成为城市的有机组成部分。

Suzhou Railway Station is located in the north of the Old City. the development of rail transportation requires a new dedicated and larger transportation center to replace the existing facilities. The design's challenge lies in ensuring that the large volume suits the small-scale traditional atmosphere of Suzhou. To retain the native architectural style, the design develops a rhombus symbol system covering the roof construction, the frames of windows and the patterns of paving. Additionally, the white walls, black tiles, gardens and corridors all ensure that this modern station becomes an organic part of Suzhou city.

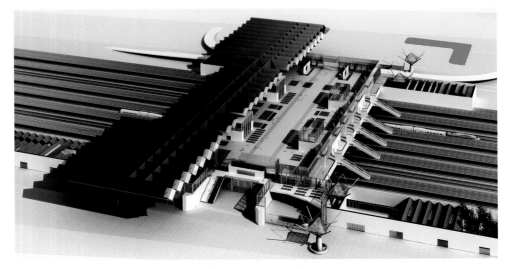

新站采取高架站房形式，南北各设入口，由高架层进站，自地下层出站。
The building adopts an elevated form. Passengers draw up at the station from the elevated level, and exit from the underground layer.

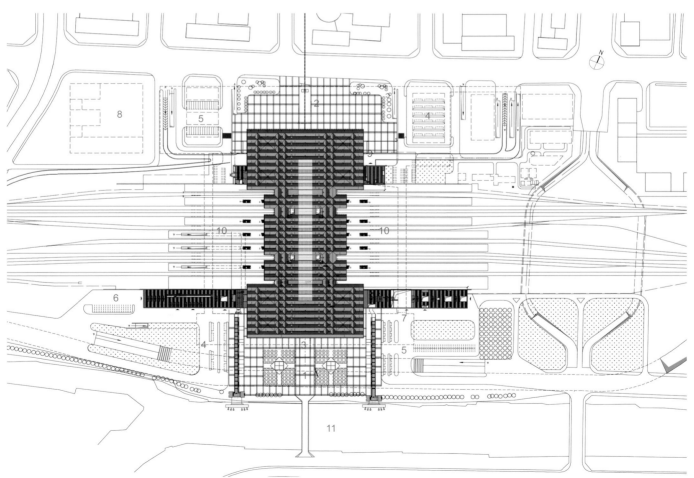

1. 南广场 2. 北广场 3. 进站口 4. 公交车停车场 5. 社会车辆停车场 6. 行包车辆停车场
7. 贵宾入口 8. 长途车停车场 9. 出站口 10. 站台 11. 护城河

总平面图

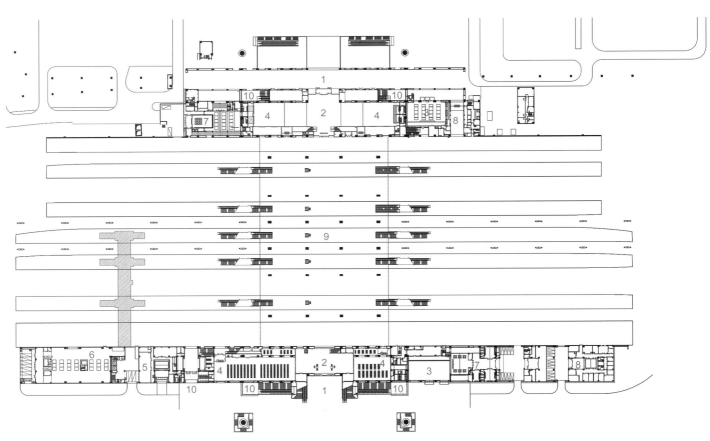

1. 进站口 2. 进站大厅 3. 售票厅 4. 候车厅 5. 办公区 6. 行包库 7. 贵宾室 8. 派出所 9. 站台 10. 出站口

首层平面图

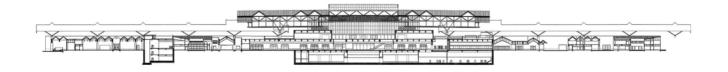

剖面图

南立面图

131

入口处屋面出檐深远，半室外的集散空间结合下沉广场、绿地、园林设置，主站房两侧围绕多个内庭院组织功能用房，营造出典雅的苏州味道的空间氛围。建筑以灰、白、栗三色为主，两组菱形灯笼柱撑起大跨度屋架，室内则采用连续的折板天花和采光窗，使菱形元素贯穿于室内外。

The cantilevered eaves beyond the entrance cover a semi-outdoor square with sunken yards and gardens. At both sides of the main entrance, a series of functions are organized around a few gardens, creating an elegant spatial atmosphere featured with a Suzhou style. Grey, white and brown composes the palette of the building, as well as the rhombus large-span roof trusses supported by the diamond-like lantern columns are highlighted indoor to deliver the sense of traditional Suzhou buildings.

庭院剖面图

苏州市吴江中学　Suzhou Wujiang Middle School

地点 江苏省苏州市 / **用地面积** 170 000m² / **建筑面积** 80 000m² / **高度** 24m / **设计时间** 2012年 / **建成时间** 2013年

方案设计	郑世伟　赵园生
	吴燕雯　白若冰
设计主持	郑世伟　莫曼春　赵园生
建　　筑	李　苗　史　倩
结　　构	崔　青
给 排 水	董　超　宋　晶
设　　备	周　锐
电　　气	许冬梅
电　　讯	王　铮　崔家玮
总　　图	齐海娟

吴江中学是一所百年老校，新校区的设计从三个方面出发：让学生在校园中感受到并珍视文化传统；注重平等、自由的学习交流；保持学生的天性。建成的校园江南园林意味浓郁，每一个庭院作为一个教学组团，使原本工厂化的教室空间被打开，与庭院空间融合在一起；同时打破班级之间的界限，通过合作教学加强交流，让教学活动变得更公开，在交流中将资讯转化为知识。每栋建筑都为学生开放，设有公共空间和服务设施，形成大量"教室以外的教学空间"，可以在校园的任何地方学习交流。造型设计秉承简洁明确的原则，体现现代建筑灵活性、流动性、匀质性等特征，同时以抽象的形式合理地表现传统意蕴。

The design of this time-honored school focus on three principles: to arouse respects for traditions, to stimulate creativity by equal communications and to keep the nature of students. Each teaching cluster is arranged to be a courtyard to make the typically introverted classrooms open to the courtyard. The boundaries of grades and classes are blurred to make the teaching activities much more public, so that students could exchange their ideas. Since most of the buildings are equipped with enough public spaces, the "learning spaces outside classroom" make it very easy to study in anywhere of the campus. The sense of traditional Chinese garden is catalyzed by abstracted traditional elements in the campus with its clean, modernist expression.

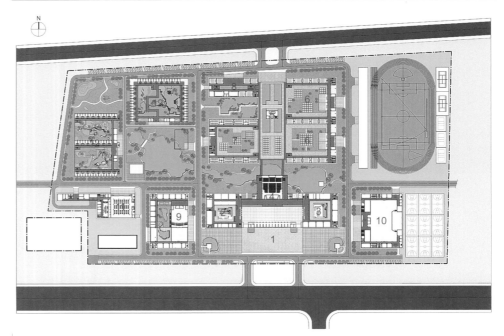

1. 主入口　2. 图文楼　3. 信息楼　4. 报告厅　5. 校史馆
6. 教学楼　7. 实验楼　8. 艺术楼　9. 行政楼　10. 体育馆　11. 食堂　12. 宿舍

总平面图

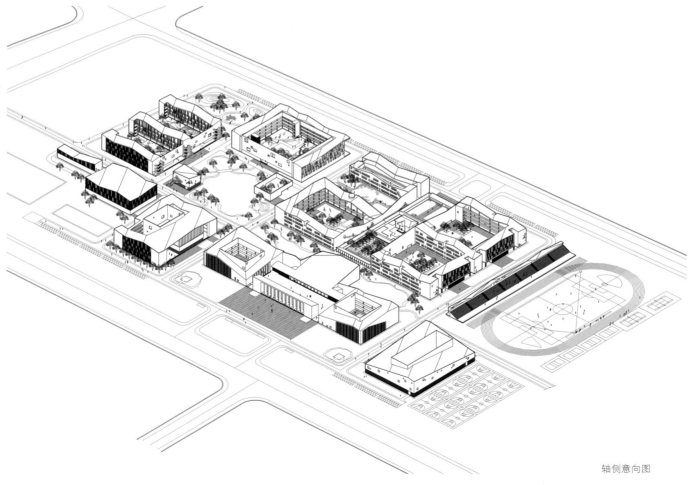

轴侧意向图

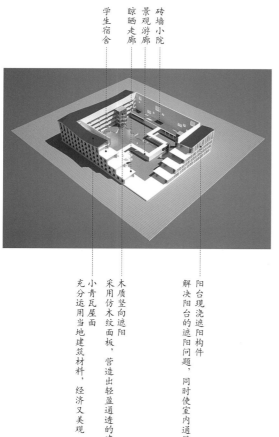

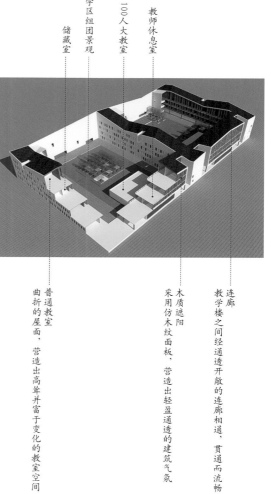

- 学生宿舍
- 晾晒走廊
- 景观游廊
- 砖墙小院

- 教师休息室
- 100人大教室
- 教学区组团景观
- 储藏室

木质竖向遮阳采用仿木纹面板，营造出轻盈通透的建筑气氛

阳台现浇遮阳构件解决阳台的遮阳问题，同时使室内通风流畅

小青瓦屋面充分运用当地建筑材料，经济又美观

普通教室曲折的屋面，营造出高耸并富于变化的教室空间

木质遮阳采用仿木纹面板，营造出轻盈通透的建筑气氛

连廊教学楼之间经通透开敞的连廊相通，贯通而流畅

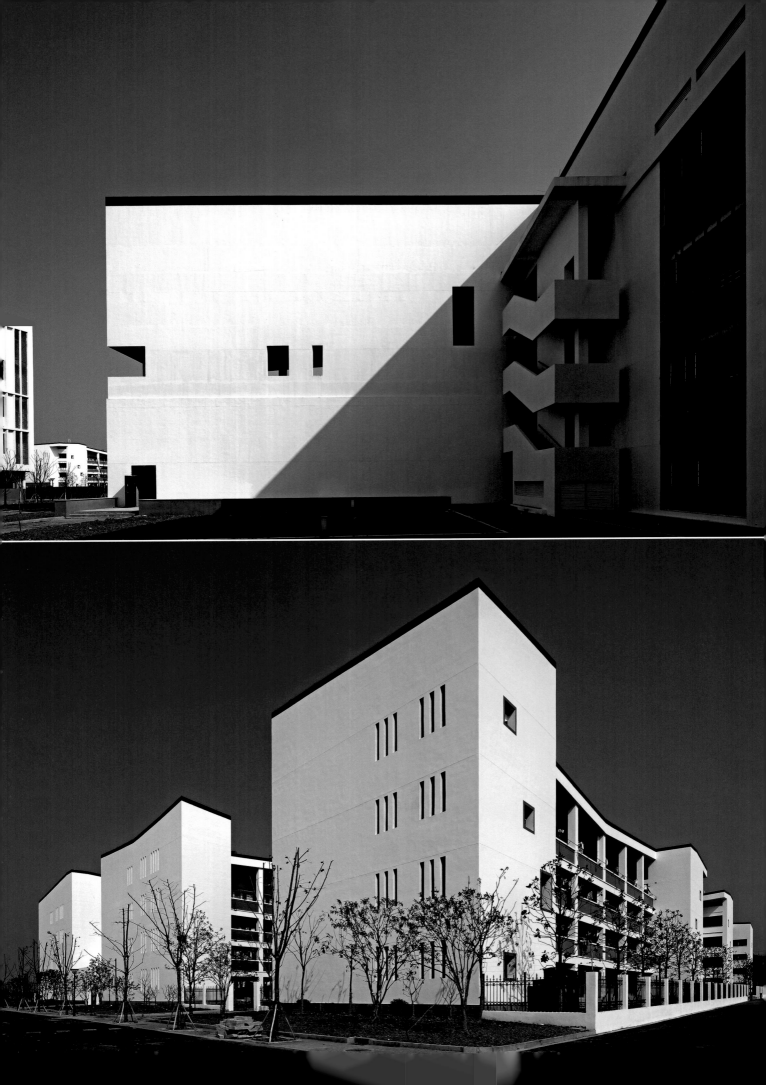

南京艺术学院改扩建项目　Renovation of Nanjing Arts Institute Campus

地点 江苏省南京市 / **建筑面积** 美术馆12 715m²、演艺大楼43 062m²、学生宿舍9 424m² / **高度** 美术馆24m、演艺中心50m、学生宿舍34m
设计时间 2008年 / **建成时间** 2012年

美术馆
方案设计 崔　愷　赵晓刚　张　男
　　　　　　高　凡　哈　成　董元铮
施工图设计 深圳华森建筑与工程设计
　　　　　　顾问有限公司南京分公司
设计主持 崔　愷　张　男　买有群
建　筑 赵晓刚　王松柏　高　凡
　　　　　张　燕　张　辉
结　构 刘先明　杨　苏　顾　建
　　　　　邓　斌　沈　伟
给排水 高　峰　张文建
设　备 邬可文　郎　平
电　气 许冬梅　王　莉　王为强

演艺大楼
方案设计 崔　愷　张　男　赵晓刚
　　　　　　董元铮　叶水清
施工图设计 深圳华森建筑与工程设计
　　　　　　顾问有限公司南京分公司
设计主持 崔　愷　时　红　买有群
建　筑 从俊伟
结　构 邵　筠　孙梓俊　董贺勋
给排水 高　峰　张文建
设　备 劳逸民　柏　文　郎　平
电　气 许冬梅　王为强
总　图 徐忠辉

学生宿舍
方案设计 崔　愷　时　红
　　　　　　张汝冰　叶水清
设计主持 崔　愷　时　红
建　筑 叶水清　熊明倩
结　构 邵　筠　尹胜兰
给排水 赵　昕
设　备 梁　琳
电　气 张　青　梁华梅　王　铮
总　图 徐忠辉

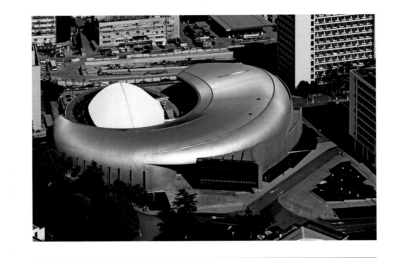

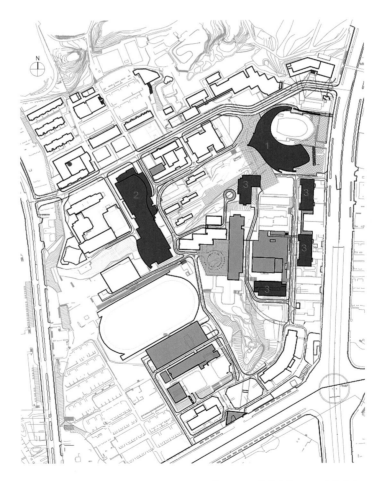

1. 美术馆　2. 演艺大楼　3. 学生宿舍

总平面图

南京艺术学院美术馆的建设基地紧南京艺术学院美术馆的建设基地紧邻原有的音乐厅。为了化解紧张用地内增加新建筑的矛盾，同时对音乐厅不完整的形式进行修复，设计柔化了美术馆的形体，以向心的弧形体量与椭圆形音乐厅扣合在一起，形成紧密的"共生体"。美术馆的机房部分被埋入地下，释放其屋面形成一个面向街道的艺术广场，为城市提供了具有凝聚力的公共空间。除了面向城市的东侧主入口，面向校园的西侧设有供师生使用的次入口，北侧则为独立的办公区。美术馆的主要展览区通过坡道连接，参观者可以在观展过程中体验不同空间的转换。疏散楼梯、通风管道等辅助性功能集中在中央的混凝土核心里，确保获得无柱的展厅。邻南京艺术学院已有的音乐厅。为了化解紧张用地内增加新建筑的矛盾，同时对音乐厅不完整的形式进行修复，设计柔化了美术馆的形体，以向心的弧形体量与椭圆形音乐厅扣合在一起，形成紧密的"共生体"。

Encircling the already built concert hall, the gallery of Nanjing Arts Institute is actually designed as a complement for the former one. It not only indicates the centripetal arc-shape is a perfect counterpart of the oval pavilion, but also emphasizes that the roof of the gallery's underground equipment rooms between them, becomes an attractive urban space and joints the two buildings cohesively. Since the east side with main entrance is designed for citizens, a campus entrance is opened on the west side as well as the north part of the building is an individual office area. Ramps connect all the exhibition spaces to provide visitors a continuous and unique experience sequence. To guarantee the column-free space effect, all the emergency staircases, ventilating pipes and other auxiliary functions are gathered into the central concrete core.

草图

向心的弧线形体对音乐厅形成半围合之势，完整、流畅、富于视觉冲击力，有助于强化识别性，确立这一公共建筑作为城市地标的特质。在经过整体规划、改造从而实现建筑与外部空间有机复合的校园空间内，这一点睛之笔起到活跃整体空间的作用，其浑然朴拙的形象也成为校园街道和广场所见精彩的视觉底景，表现出美术馆独特的艺术感染力。

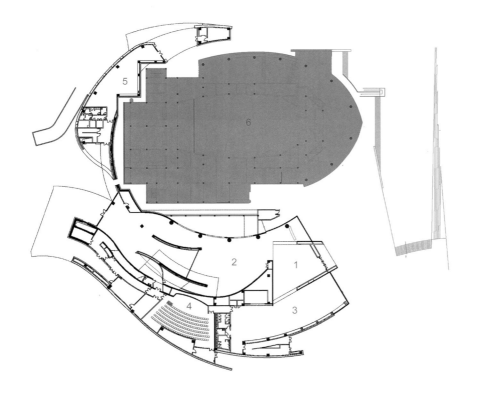

1. 门厅
2. 展厅
3. 临时展厅
4. 报告厅
5. 咖啡厅
6. 原有音乐厅

美术馆首层平面图

Effortlessly, the building turns into a focal point of the community by its remarkable fluent figure. For the organically organized campus, it also acts as a catalyst for activating the whole space structure. Its powerful image becomes the end of the view corridors of public spaces in the campus.

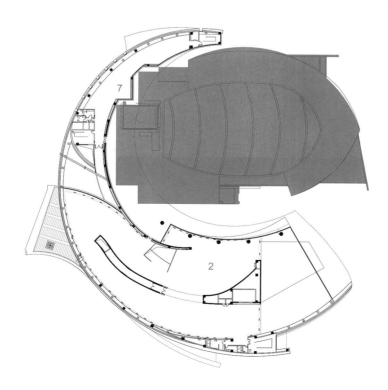

7. 校史馆

美术馆三层平面图

摄影 王彦

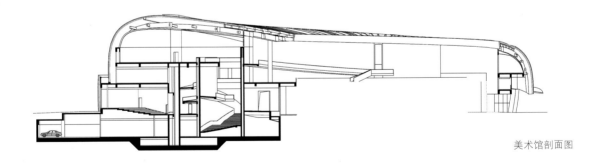

美术馆剖面图

演艺大楼位于校园一处狭长的南北向用地上，地势高差约12m，周边房屋密集。其形态来自设计应对环境的策略，与现状建筑在尺度和空间上寻找关系并实现了协调。在满足功能和面积需求的前提下，设计尽量限制地面以上的建筑体量。结合地势将不同功能的入口设于不同楼层上，所形成的入口广场同时承担了衔接地势，加强南北校区联系的功能。为避免过长的建筑体量成为东西向视野的阻挡，建筑中部断开并加以削减，在三层入口形成景观视线通廊。横向坡道形成环廊，解决了高差问题，并连通各层内廊，提高了走廊使用率。所有练琴房均有自然采光，不需要自然光的大空间排演厅等则设于平面中部及地下。建筑立面是对内部空间的真实反映。东西向外廊采用成排通透的百页阵列；面宽不一的外挂琴房使得立面开窗和空调室外机格栅产生跳跃的节奏感。格栅、开敞平台与半敞开外廊等形成了既有规律又充满变化的组合，并与剧场区大面积的清水混凝土墙面形成对比。

As an insertion to such a dense built area, the performance building is considered to be a coordinator for the existing buildings and a connection for two originally separated campuses. Building configuration derives from the strategy answering its environment in scale and in reconstruction of public spaces. Under the promise to satisfy functional requirements, the mass above ground is minimized. The 12m height difference of the narrow site is exploited for disposing entries of different functions on different levels so that to shape a few entrance squares for the building itself and its neighbors. To avoid blocking the east-west-direction views in the campus by such a long building, the central part is cut down. Ramps, stairs and platforms form a outside loop system to accommodate the complicated height difference. Every music room owns natural lighting, while those large scale rooms for rehearse or performance are set in the central of plan or underground. The appearance exactly represents the space organization. The outside loops are wrapped by louver boards and the different widths of rooms give the façades a rhythmic atmosphere. In contrast to the large solid concrete wall of auditorium, the louvers, platform and side corridors produce multi-layered effects.

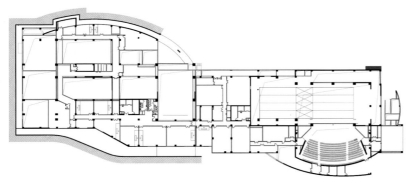

演艺大楼剧场池座层平面图

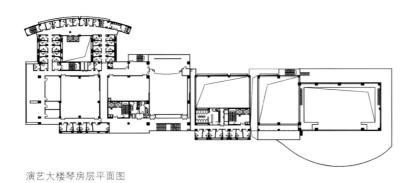

演艺大楼琴房层平面图

供四个学院使用的排练厅、演奏厅、观演厅和琴房对高度、面积的要求各不相同,要在有限的体量内塞进数量和种类繁多的房间,经过错综复杂的排列组合,以平面上6m、11.2m和8.4m的三联柱跨,配合2.6m、3.9m、4.2m、7.8(3.9×2)m和8.4(4.2×2)m五种层高的复杂搭配,得到了从4m²到400m²不等的近600间琴房和上百个无柱的排练、表演和教学房间。

After a painstaking arrangement process, more than 600 music rooms, a lot of rehearse halls and an auditorium is filled into the limited volume. To satisfy the different required sizes, heights and positions, the spans of the plan are decided to be, in order, 2.6m, 3.9m and 4.2m, with 5 different floor heights, 2.6m, 3.9m, 4.2m, 7.8 (3.9×2) m and 8.4 (4.2×2)m.

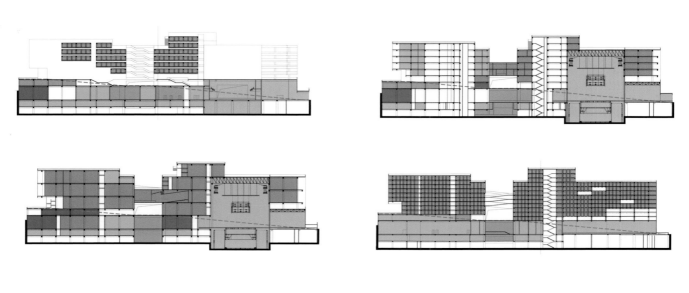

■ 琴房　　■ 舞蹈学院　　■ 音乐学院　　■ 流行音乐学院　　■ 高职学院　　■ 剧场　　　　　　　　剖面示意图

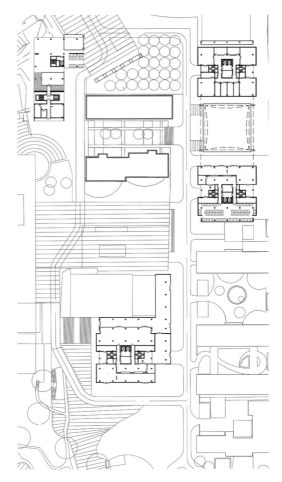

为满足学生宿舍的需求量，在狭小用地内新建的四栋宿舍楼均为板式高层，它们共同围合出内聚性的空间，并通过保留原有大树，底层架空，减少对环境的压迫感。架空层内设有茶吧、洗衣房、管理室、小超市等服务功能，地下为半开敞的自行车库。通透的廊道围合出庭院，也成为通往校园公共空间立体步行系统的一部分。宿舍外错落的公共平台创造了更多交往的机会，外阳台和两间共用一套的卫生间则确保了学生的生活质量。

To accommodate a large number of students, four students' dormitories are inserted into a narrow slope. They enclose the quadrangle, providing place for tea bar, laundry room, management office and small supermarket on their semi-opened bottom floor. The twisting colonnade between them also becomes a part of the whole pedestrian system of the campus. Platforms on each floor also act as the communication space for students. In addition, each dormitory has a balcony and clotheshorse and two dormitories share a toilet, for the purpose of improving the life quality of students.

宿舍首层平面图

北京外国语大学图书馆新馆 New Library of Beijing Foreign Studies University

地点 北京市海淀区 / 用地面积 4 500m² / 建筑面积 24 292m² / 高度 24m / 设计时间 2006年 / 建成时间 2013年

设计机构	中旭建筑设计有限责任公司
方案设计	崔愷 王祎 潘悦 辛钰
设计主持	崔愷 张念越
建 筑	张念越 辛钰
结 构	许忠芹
给 排 水	张之川 李严
设 备	郭晓楠
电 气	姜晓先
总 图	夏菌颖
室 内	刘烨 张晔 饶劢

学校原有的图书馆建于20世纪90年代，已不能满足当前使用要求。由于用地紧张，加建环绕原有建筑四周展开，最大限度提高了土地使用效率，增加了阅览、藏书面积。开放式藏阅合一的空间，更为适应当前阅览要求。庭院经过改造成为退台式的共享中庭，南北两侧为开放式阅览，新老建筑交接处为交通空间。设计部分保留了旧有建筑，并综合考虑周边建筑元素。紧邻人行道路一侧的首层立面采用大面积通透的玻璃窗，减少行人的压迫感。西侧结合突出的报告厅将底层架空，便于学生穿行，使建筑尽量融入校园。二层以上立面以虚实结合的形式演绎了"书架"的概念。多种语言的"图书馆"拼写而成的外墙，也成为对外国语大学特色的强调。

The original library was built in early 1990s and cannot meet the current functional requirements. Within a limit site, the design preserves the original structure, and increases floor area by making extension to the surrounding places, that creates an open space for book collection and reading. The courtyard is devised to be a stepped back atrium, which is flanked by two open reading areas. The elements of original and surrounding buildings are integrated into the new one. By maintain a passage to the playground by opening the bottom on west; it infuses itself into the campus in a sensitive way. For the design of façades, besides its imitation of "book shelf", the most distinguished is the GRC-precut wall, which shows the word "library" in more than 50 different languages, becoming a visual icon of the university.

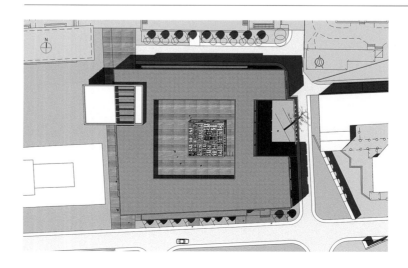

总平面图

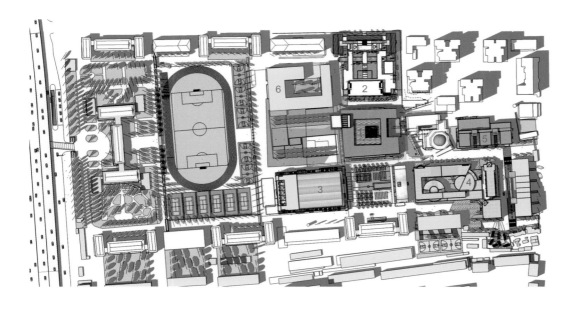

1. 图书馆
2. 行政楼
3. 体院馆
4. 中国语言文学学院
5. 逸夫楼
6. 英语中心

位置图

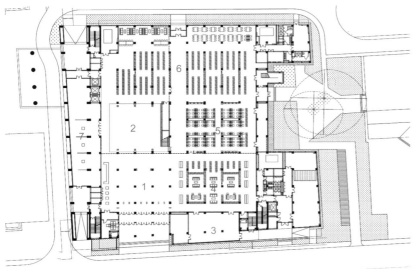

1. 门厅
2. 共享大厅
3. 咖啡厅
4. 休闲阅读区
5. 电子阅览区
6. 报刊阅览区
7. 展厅

首层平面图

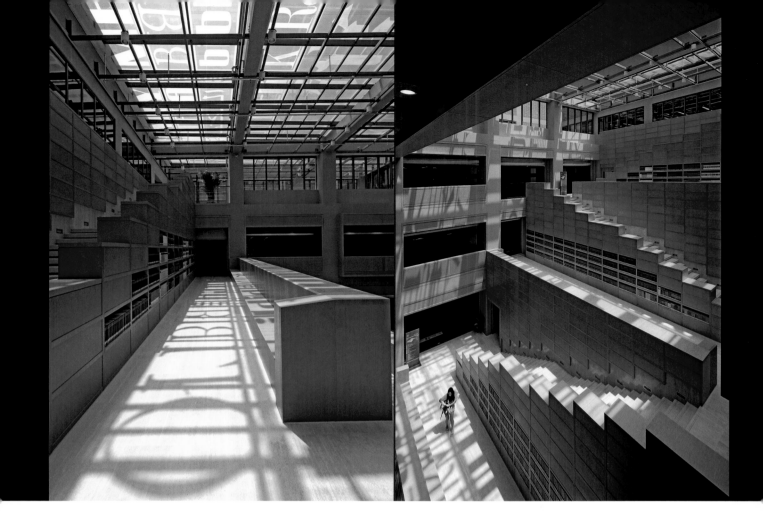

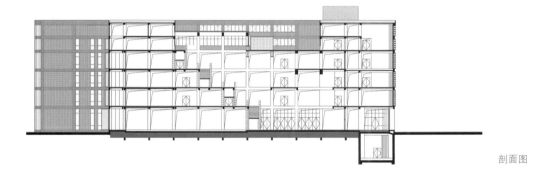

剖面图

北京华文学院新校区 New Campus of Beijing Chinese Language and Culture College

地点 北京市昌平区 / 用地面积 145 771m² / 建筑面积 128 353m² / 高度 26m / 设计时间 2009年 / 建成时间 2012年

方案设计 崔海东 王敬先
设计主持 崔海东 王敬先
建　筑 饶洁 靳树春
结　构 张付奎
给 排 水 黎松
设　备 宋玫
电　气 王亚冬 阚璇
总　图 刘晓琳

北京华文学院的新校区可以容纳1500名学生，其中教学行政区位于校园的中心地带，是学校的功能主体；宿舍区设于安静的东北角。食堂、体育馆、游泳馆等三个大体量的建筑则紧邻体育场布置。教学办公楼作为校区核心建筑位于南北向中轴线上，其南侧左右并置两座行政办公楼，会堂和图书馆设于北侧，与主楼东西两翼的实验教室区域相连，使得学生通过室内通廊就可以到达所有教学用房。同时，图书馆的设置也靠近宿舍区，便于学生使用。立面以西洋古典风格为基调并加以简化，外墙装饰中运用中华纹样，在体现国际校园特色的同时反映中华传统文化。

A new campus for Chinese language teaching, the project includes a teaching & administrative building, dormitories, a dining hall, a gymnasium and a natatorium. The teaching & administrative building, as the centerpiece of the campus, is located on the north-south axis. On the south of it are two administrative wings, while the audience hall and library are sited on the north and connected to the laboratory classroom areas of the main building. The location of library is very close to the dormitories to facilitate students' accesses. Simplified western classical vocabulary and Chinese style ornaments on façades reveal the nature of the college, an international school for Chinese culture promotion.

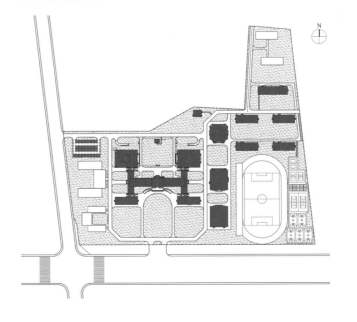

1. 教学楼
2. 办公楼
3. 图书馆
4. 报告厅
5. 游泳馆
6. 体育馆
7. 食堂
8. 学生宿舍

总平面图

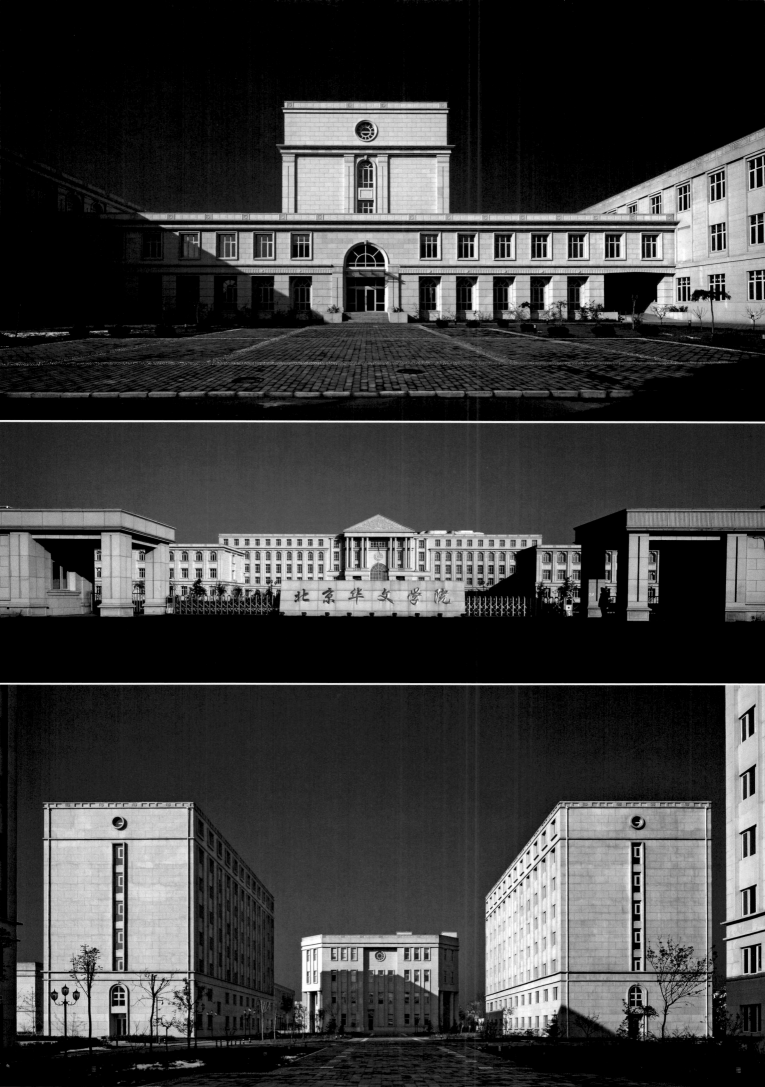

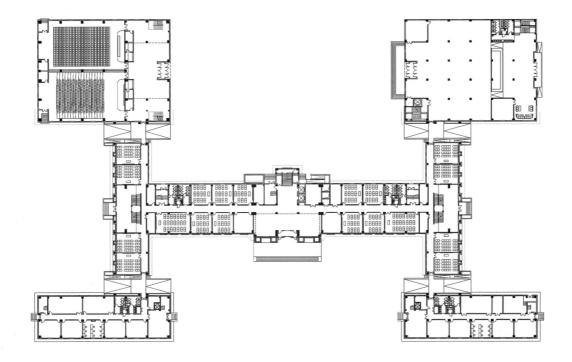

教学行政区首层平面图

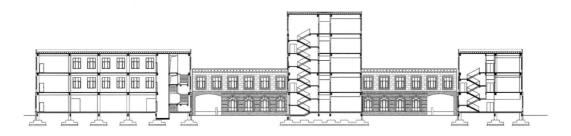

教学行政区剖面图

德阳市奥林匹克后备人才学校 Deyang Olympic Sports School

地点 四川省德阳市 / 用地面积 100 942m² / 建筑面积 17 422m² / 高度 20m / 设计时间 2009年 / 建成时间 2012年

方案设计	崔愷 关飞 傅晓铭
设计主持	崔愷
建 筑	关飞 傅晓铭 彭书明
结 构	施泓 鲍晨泳
给 排 水	石小飞
设 备	宋玫
电 气	陈琪 王烈
总 图	王炜

该项目是由国际奥委会、中国奥委会、北京奥组委在汶川地震后捐助800万美元建设的。设计思路是将传统的体校设计转换成开放式的城市体育公园。十字形的建筑把室外运动场地分为田径、篮/排球和网球3个区，校舍、球馆、泳池、网球场在十字平面内沿4个方向展开。中心是更衣沐浴和管理用房，相互之间以休息长廊连接。建筑结合德阳的气候特点，尽量采用半开放、自然通风采光的空间，以利节能。建筑材料主要为结构性清水混凝土。钢拱屋面和本地灰砖、竹材，表达了健康、生态的理念，强化了建构的逻辑，也从某种意义上体现了奥林匹克精神。

The project is funded by the International Olympic Committee and Chinese Olympic Committee after the 2008 Wenchuan Earthquake. The cross-shaped building creates outdoor sports venues for athletics, basketball/volleyball and tennis. It also distributes the school's dormitory, arena, swimming pool and tennis courts in four directions. Spaces for bathing, dressing and management are located at the center and connected by long corridors. The building is adapted to local climate and characterized by semi-open space with natural ventilation and lighting. The dominant material, exposed structural concrete, incorporated with bowed steel roof, gray bricks and bamboo, expresses intentions about sustainability.

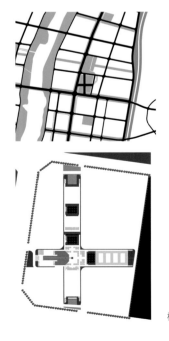
概念图

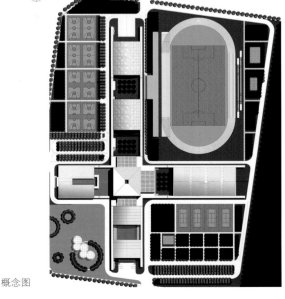
总平面图

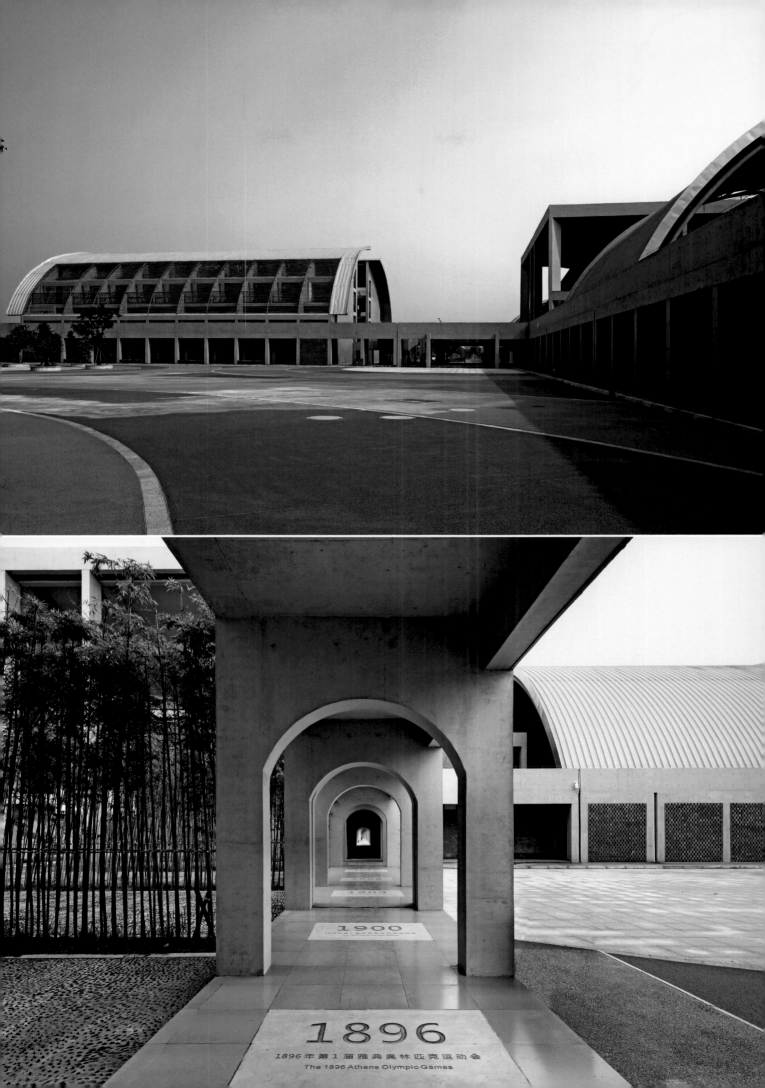

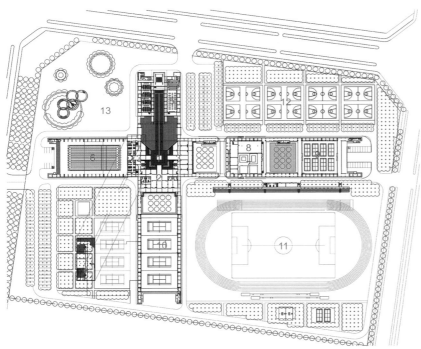

首层平面图

1. 共享区
2. 接待区
3. 辅助用房
4. 教室
5. 餐厅
6. 室内游泳池
7. 重竞技馆
8. 健身馆
9. 羽毛球馆
10. 网球场
11. 田径场
12. 篮球场
13. 奥林匹克广场

剖面图

鄂尔多斯东胜图书馆　Erdos Dongsheng Library

地点 内蒙古鄂尔多斯市 ／ **用地面积** 45 600m² ／ **建筑面积** 32 031m² ／ **高度** 23m ／ **设计时间** 2010年 ／ **建成时间** 2012年

方案设计	崔　愷　柴培根
	朱　迪　田海鸥
设计主持	崔　愷　柴培根
建　筑	朱　迪　蒋　鑫
结　构	段永飞
给排水	侯远见
设　备	宋　玫
电　气	王亚冬
总　图	徐忠辉　邵守团
室　内	张　晔

东胜图书馆包括图书馆和规划展览馆两部分功能。建筑位于城市新区的公园内，一侧面向城市主要道路，为了平衡城市和公园两种尺度的不同要求，沿城市街道的外侧用连续的界面迎合道路转合的趋势，面向公园的内侧拆解体量以适应公园的尺度。表现在形式上，则以内蒙古的区花"马兰花"作为形象，如花瓣般错落层叠的阅览区，面向公园，通过退台使景观和建筑更紧密地融合在一起。朝向道路的立面采用连贯平滑的横向金属铝板，入口门厅处采用通透的玻璃幕墙，虚实立面相互穿插，提供了既完整又生动的城市界面。

Housing a library and a city planning exhibition hall, the complex is situated between a peaceful park in the newly-built district and a main road of the city. The bold design resolves the site's contradictory nature by juxtaposing a series of curvilinear volumes that bend inward. By facing the road with their continuous surfaces, the ends of them arrayed perpendicularly to the park give the building a small scale appropriate to integrate into the scale. The composition of the curves also features as the province flower of Inner Mongolia, Malan Flower. Meanwhile, the terraced platforms outside the reading rooms provide readers opportunity to enjoy the landscape.

草图

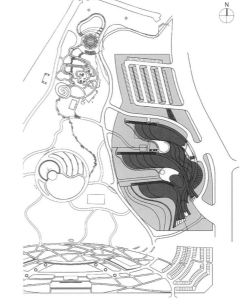

总平面图

曲线的建筑轮廓，使内部空间流动感更加强烈，为了不破坏空间的流动性，结构取消了梁柱的常规做法，采用曲线片墙作为支撑体系，使建筑内部空间显得更加连贯。

Aim to rendering a continuous and diverse inner space, the reading areas are modeling like petals of the flower. Curved wall as the whole structure system which replaced regular beam column system, also do much contribution to express the continuity of the space.

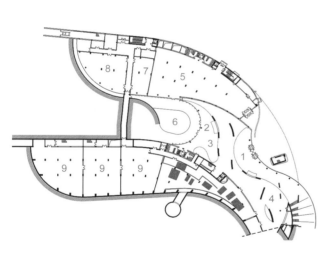

1. 门厅 2. 检索 3. 休息 4. 电子阅览 5. 儿童阅览
6. 室外庭院 7. 编目 8. 采编 9. 密集书库 10. 阅览室

−4.800m入口层平面图

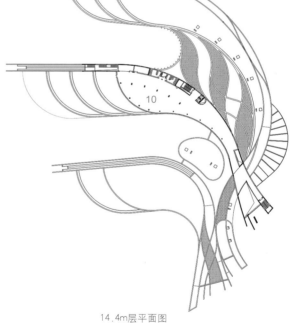

14.4m层平面图

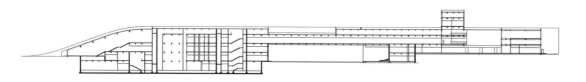

剖面图

万州体育中心 Wanzhou Sports Center

地点 重庆市万州区 / 用地面积 187 000m² / 建筑面积 49 704m² / 高度 65m / 设计时间 2009年 / 建成时间 2012年 / 座位数 体育场 26 000座

方案设计	李燕云 罗 洋 赵梓藤 王 斌 关 帅
设计主持	李燕云 戴 星
建　筑	罗 洋 赵梓藤 关 帅
结　构	刘松华 唐 杰
给排水	朱跃云
设　备	李京沙 马 豫
电　气	王玉卿 王浩然 林 佳
总　图	王亚萍 李可溯
景　观	冯 君

万州体育中心东临长江，面朝青山，其设计与环境充分结合。主体育场与游泳馆呈"八"字形布置，形成环抱江景之势。体育场看台采用了不对称的形式，沿江一侧较小且不设罩棚，让主看台的观众在比赛同时也能欣赏到江景山色，巧妙地将场外的山水与场内的活动交融于一体。主看台罩棚的落地钢管桁架结构，跨度250m，成为体育中心最为突出的元素。呈折扇状的罩棚犹如江上张开的白色巨帆，与青山绿水相映成趣。沿江看台后大面积的绿坡也削弱了庞大体量的视觉影响。

Nestled at the foot of mountains and facing Yangzi River, the location of the sports center request a design to with the environment adequately. The design masses a relatively small stand on the waterfront to allow views from the main stand to the river. It features interactive displays of human activities and natural environment. With a span more than 250m, the steel truss supporting the main canopy creates a signature identity of the site. Since most of the concrete volume is hided under grass slop, the white canopy, inspired by the sailing ships on Yangzi River, figures a poetic image in the nature.

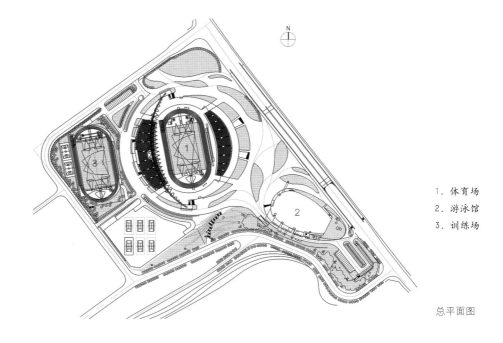

1. 体育场
2. 游泳馆
3. 训练场

总平面图

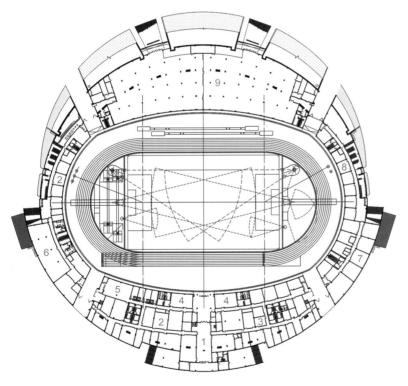

1. VIP入口
2. 办公
3. 新闻中心
4. 运动员休息
5. 检录处
6. 健身中心
7. KTV
8. 精品店
9. 商场

体育场首层平面图

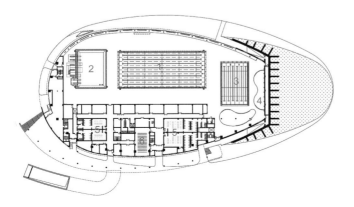

1. 比赛池
2. 跳水池
3. 训练池
4. 戏水池
5. 更衣、淋浴
6. 新闻中心

游泳馆首层平面图

游泳馆主体呈椭球形,屋面、墙面浑然一体,造型有如一颗江边晶莹的水珠。与主体育场类似,其南侧也嵌入绿坡地,延续了从自然中生长出来的形态。室内功能布局从引入江景的角度出发,跳水、游泳比赛厅沿长江一字排开,且沿江一侧完全通透,在比赛的同时可以尽享外部景观。

The natatorium's exterior is dominated by an asymmetrical ovoid roof that appears as a drop of water. Similarly, it is also embedded in a sloping lawn and seemingly emerges from the earth. To strengthen the natatorium's relationship with the nature, the diving pool and swimming pools occupy the side along the transparent glass curtain wall. So the audiences and swimmers will enjoy panoramic views of Yangzi River.

昌吉体育馆   Changji Stadium

地点 新疆昌吉市 / **用地面积** 34 138m² / **建筑面积** 16 385m² / **高度** 27m / **设计时间** 2010年 / **建成时间** 2013年 / **座位数** 体育场 26 000座

方案设计	曹晓昕 孙雷 娄阁
设计主持	曹晓昕 孙雷
建　　筑	娄阁 周龙
结　　构	史杰
给 排 水	周博
设　　备	刘筱屏
电　　气	徐世宇
总　　图	连荔
室　　内	邓雪映

昌吉体育馆的设计理念是将若干个相似的散落个体，按照一定的秩序紧密地结合在一起，组成一个充满力量感、无法分割的整体。形式的生成方式来源于对伊斯兰传统建筑特征的理解和演绎，而整体状态又契合体育馆对大空间的需求。经过整理和提炼民族地域性的建筑符号和细节，以现代的手法融入建筑的细节处理之中，使建筑具备了比较强的可识别性。外立面材料的运用，通过将同一种石材进行两种不同的加工处理，使石材在不同光线下表现出个体差异。其菱形的图案，同样来自于对地域特色的考虑。

The matrix of individual square modules gestures as a powerful and indivisible entity that follows a certain geometric order. It reinterpreted the traditional pattern of Islam architecture in a modern manner as well as creates large-span spaces to satisfy the requirements of sports games. Detailing selected and abstracted from the symbols on local Uyghur buildings, creates an indigenous feature for the native Uyghur people. The subtle contrast of the rough and smooth texture on each granite panel creates ample details under the sunlight, while the diamond-shaped pattern is also referenced from the native buildings.

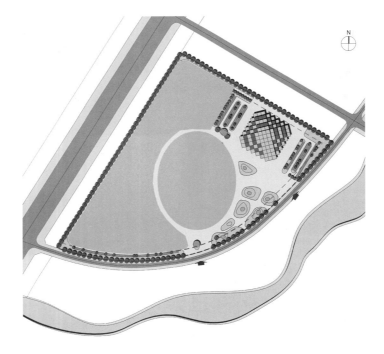

总平面图

剖面图

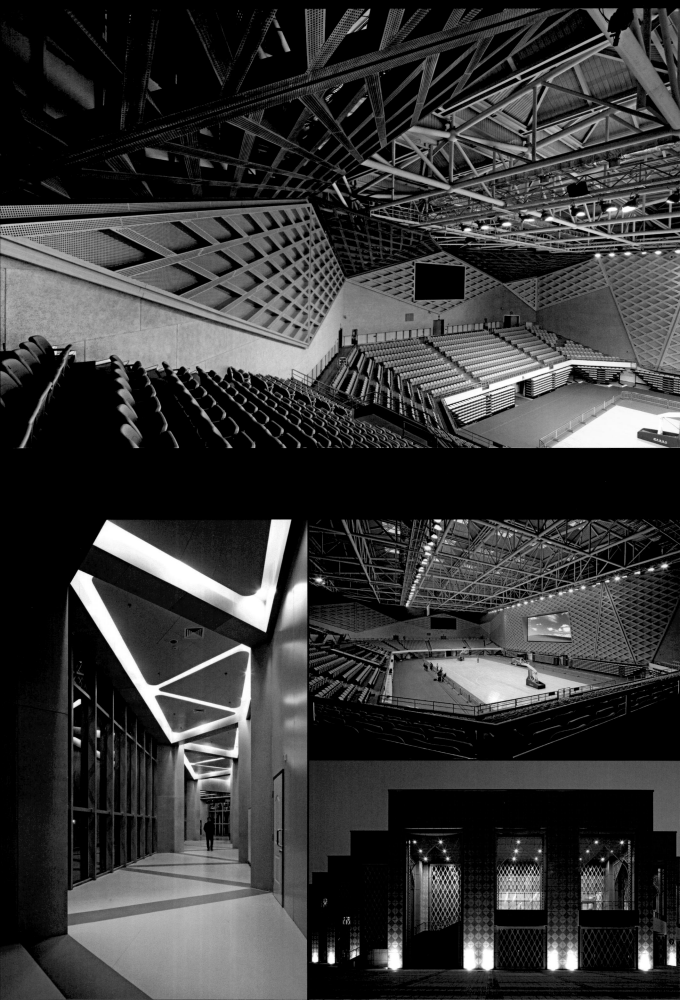

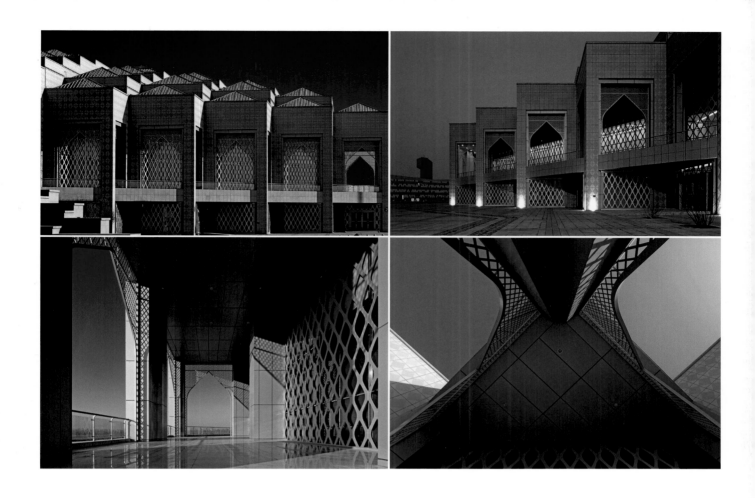

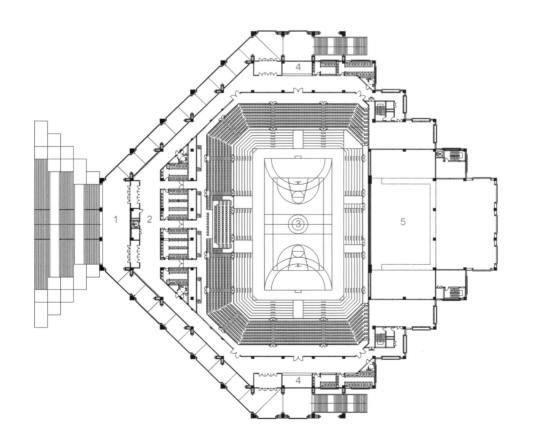

1. 外廊
2. 观众集散厅
3. 比赛大厅
4. 商店
5. 训练厅上空

二层平面图

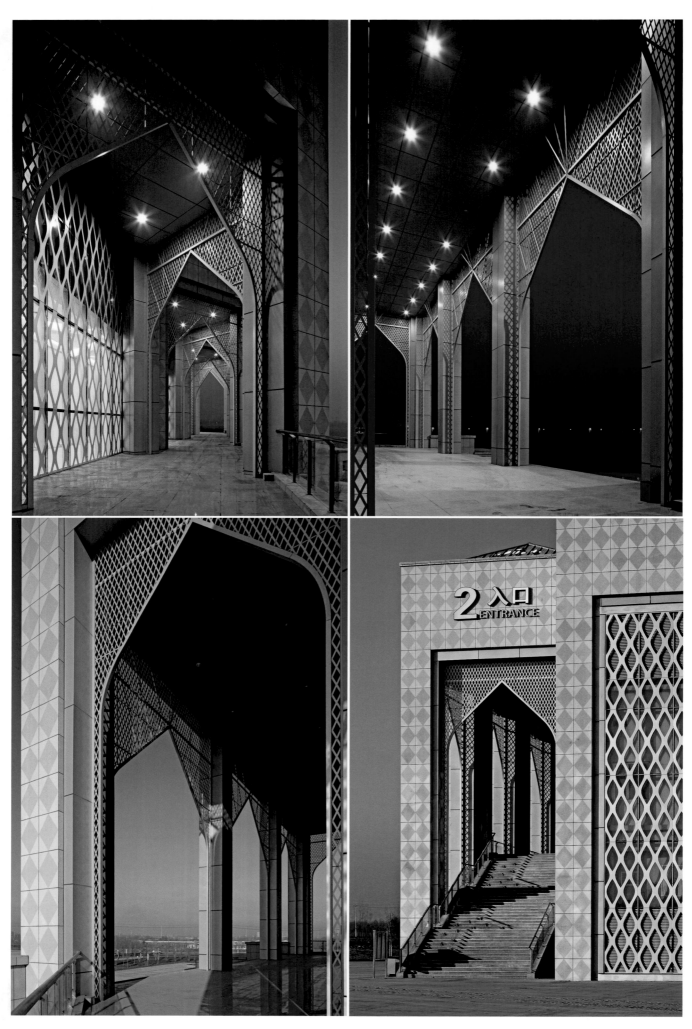

北工大软件园二期E地块 BPU Software Park II-Plot E

地点 北京市亦庄经济开发区 / 用地面积 41 337m² / 建筑面积 77 504m² / 高度 43m / 设计时间 2010年 / 建成时间 2012年

方案设计　崔　愷　柴培根
　　　　　周　凯　任　玥
设计主持　崔　愷　柴培根
建　　筑　周　凯　任　玥　杨文斌
结　　构　毕　磊　周　昕
给 排 水　王耀堂
设　　备　李京沙
电　　气　王浩然
电　　讯　许士骅
总　　图　连　荔
景　　观　冯　君

地块位于本院此前设计的B地块的西南侧，西面为风景优美的凉水河。设计充分利用这一景观资源，将较高的建筑布置在沿河一侧以获得良好的景观，同时建筑间较大的日照间距也确保园区内部可以欣赏到河畔景色。建立与软件园其他地块的联系是设计的另一要点。三层高的裙房将高层建筑连接起来，这个高度与A地块现有建筑相同，便于形成完整的城市界面，使整个区域具有识别性。顺应弧形道路所形成的折线层叠变化，既提示了道路特点，也增加了建筑的标志性。为园区提供服务的企业会所，设于地块东北侧，与B地块的公共服务平台联系紧密，增加了各区之间的联系。

Located along Liangshui River, the Plot E of BPU Software Park II is just to the southwest of the former built Plot B, also designed by CAG. By setting the higher office buildings facing the river, it gains larger sunshine spacing to introduce more scenic views into the deep of the campus. Another key to the design is the continuous street wall paralleling the riverbank, which creates a signature identity, while respecting the context with the same height of the podium. The enterprise club of the campus is on the northeast corner, which is close to the public service building of the Plot B, to intensify its connection to other campus of the whole business park.

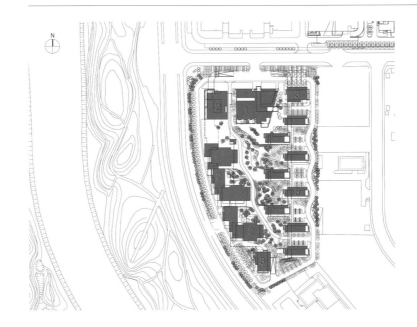

1. 企业会所
2. 集约办公
3. 独栋办公

总平面图

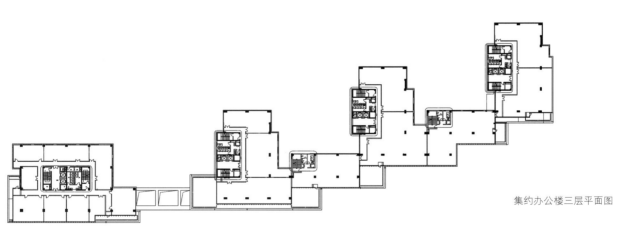

集约办公楼三层平面图

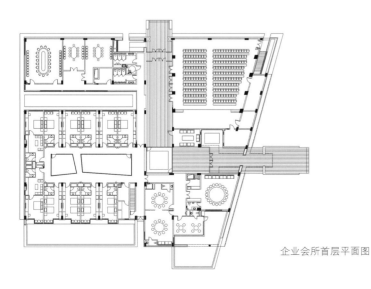

企业会所首层平面图

中国科学院电子学研究所怀柔园区 Huairou Park of Electronics Institute, Chinese Academy of Sciences

地点 北京市怀柔区 / 用地面积 67 496m² / 建筑面积 74 266m² / 高度 36m / 设计时间 2009年 / 建成时间 2013年

方案设计 逢国伟 王超若 孟可
设计主持 逢国伟 王超若
建　　筑 苗轶默 任肖莉
结　　构 张根俞 席志刚 宋力
给 排 水 石小飞
设　　备 邬可文
电　　气 许东梅
总　　图 邵守团

本科研园区的布局，采用一中心带四组团的向心式格局。园区中心的办公建筑统领园区其他建筑，同时也作为面向城市的主要建筑形象。单体建筑结合研发试验工艺要求，采用四个相似的U形平面，组织出若干个私密或公共的庭院空间，强调建筑间的围合感与秩序感，并为每个研发部门营造相对独立的景观和休憩空间。作为典型的科研实验建筑，单体建筑由通用实验室、研发空间及辅助交流空间等基本要素组成，而且具备净化、屏蔽、避震、高温等特殊功能实验室，在科技产业园类建筑中有的较高代表性和通用性。

For the scientific office park, a master plan composed of a central building and four individual modules is devised to form a remarkable skyline. The U-shape plan of the research module is derived from the technological requirements of experiments. By enclosing a landscaped court, it provides each researching department an independent exterior space. As a prototype of scientific research facility, each building consists of certain basic elements, such as laboratory, research office and communication space and several specific laboratories with functions of purification, electro-magnetic shielding, avoiding vibration and high temperature heating.

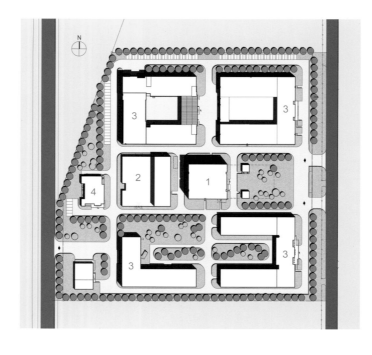

1. 设计办公楼
2. 机械加工中心
3. 研发实验楼
4. 食堂

总平面图

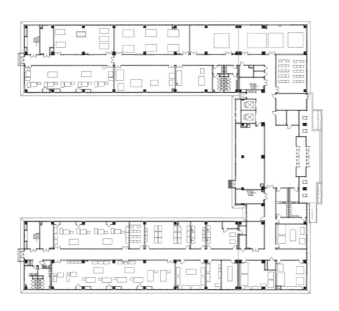

研发实验楼首层平面图

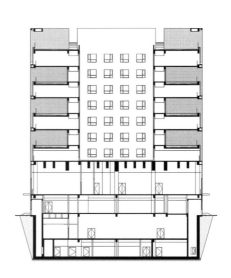

设计办公楼剖面图

绿色建筑材料国家重点实验室 State Key Laboratory of Green Building Materials

地点 北京市朝阳区 / 用地面积 1 124m² / 建筑面积 6 563m² / 高度 18m / 设计时间 2009年 / 建成时间 2010年

方案设计　杨金鹏　刘　德　闫小兵
设计主持　杨金鹏　陈帅飞
建　　筑　刘　德
结　　构　陈文渊　周方伟
给 排 水　王则慧
设　　备　李超英
电　　气　李战赠
总　　图　高　治

绿色建筑材料国家重点实验室，是中国建筑材料科学研究总院设置的专门进行绿色建材研究的机构，所研制的混凝土条形挂板，具有板材宽度小、长度可控、型材变形小、板缝可调可控等优点。其骨料大量利用建筑垃圾、粉煤灰、矿渣等固体废弃物，符合节能环保要求。在办公楼的建造中，设计者充分利用了其研究成果，将混凝土条形挂板作为设计的主要元素和出发点，使这种原本为抗震节能型乡镇住宅研发的水泥制品，在公共建筑上得到应用，并结合框架结构形式进行调整，提出了示范性技术措施，为产品的批量化生产、规范化使用和市场化开发提供了技术支持和形象展示。设计结合建筑的外立面不同的进退关系，交错使用不同的混凝土再生材料，充分展示了材料的性能。

The lab is a particular organization for the research of green building material. The concrete strip-type plate they invented is a promising material, with advantages of smaller width, controllable length, little deformation and changeable gap width between plates. A wide range of solid waste can be used as the aggregate of this type of plate, such as the construction waste, fly ash and slag. Design of the lab's office building exploits the talent of the material and makes full uses of it on a large scale public building. Before that, it was only thought an energy-saving material for vernacular houses. A series of demonstrative techniques are proposed to provide technical support and application example for the future quantity production, standardized application and market development. Fabricated by different aggregates, the plates feature characters of different building components.

混凝土条形挂板宽165mm，厚40mm，内部配筋以加强稳定性，减小板材和墙体自重。建筑入口处的板材骨料适当添加回收的贝壳、瓷片，使板材中浅灰色基调下增添独特的斑斓色彩，随阳光照射而获得生动的反光。

Placement of reinforcement in the 165mm wide and 40mm thick concrete plate could strengthen its stability and reduce its self-weight. By adding recycling ceramic chips and shells into the aggregates, the plate could glitter subtly in the sunshine.

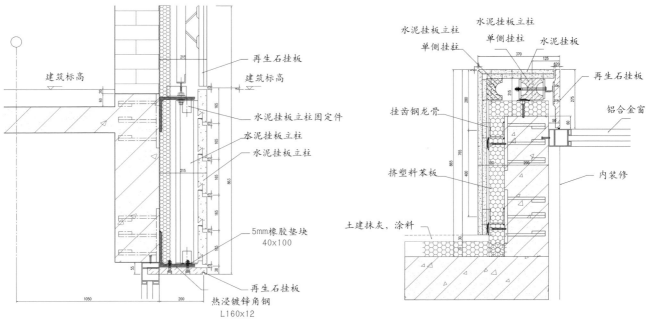

条形挂板安装标准节点图　　　　　　　　　条形挂板安装转角处节点图

巴彦淖尔市临河区行政中心　Administrative Service Center of Linhe District, Bayan Nur City

地点 内蒙古巴彦淖尔市 / 用地面积 65 420m² / 建筑面积 41 277m² / 高度 38m / 设计时间 2008年 / 建成时间 2012年

方案设计　曹晓昕　李衣言
设计主持　曹晓昕
建　　筑　李衣言
结　　构　王　载
给 排 水　匡　杰
设　　备　李雯筠
电　　气　蒋佃刚
总　　图　连　荔

办公楼独特而极富张力的外部形态，来源于行政办公建筑功能的需求，和对内蒙古当地文化的理解。建筑由入口空间开始的被挤压状态，自下而上扩散至顶层，并且在端部以同样的方式强化。建筑所呈现出的逐渐生长而蓄势待发的张力，契合了内蒙古人内心坚毅、厚积薄发的性格特征。同时，这种单元式的小开间模式，也是满足政府工作人员办公需求的结果。中庭空间因外部形态的变化而被立体分割为三个部分，墙面同样延续了外部形态的挤压状态。这使得建筑张力的传递由外及内，从整体上体现了蓄势待发的状态。

The dynamic appearance of this office building, which looks like a stack of office cells, resulted from the requirements of small width private offices in government office building. Extruded from the main entrance, the complete building embodies this imaginary motion: cells are stepped back gradually from the bottom to the top, from the center to both ends. The intensified gesture is conceived matching to the mix of tenacity and enthusiasm of Inner Mongolia people. Even the atrium is extruded into three parts, while each of them continues the tension by its terraces and stepped back floors. From outside to inside, the extrusion mode becomes an overall theme.

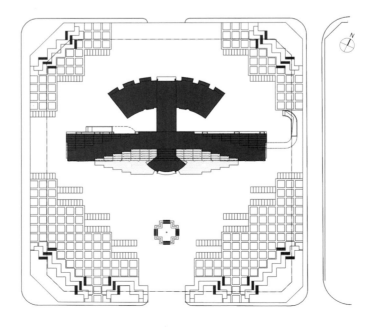

总平面图

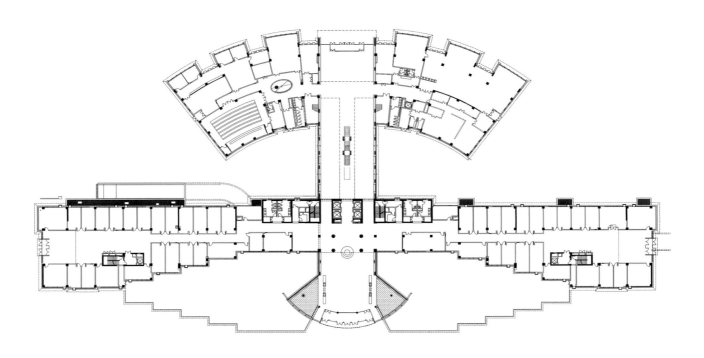

首层平面图

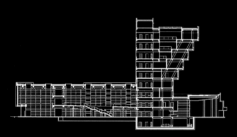

剖面图

巴彦淖尔市政务服务审批中心　Administrative Approval Service Center of Bayan Nur City

地点 内蒙古巴彦淖尔市 / 用地面积 15 377m² / 建筑面积 24 998m² / 高度 30m / 设计时间 2008年 / 建成时间 2013年

方案设计	曹晓昕　余　浩　王　帆
设计主持	曹晓昕
建　　筑	刘崇霄
结　　构	刘兴国
给 排 水	王则慧
设　　备	李雯筠
电　　气	王　莉
总　　图	王　曦

针对这一市政服务型的办公建筑，设计以"便民服务广场"为理念，强调开放、亲和、便民、高效的原则。建筑形体采用了6个圆柱体相互交错的组合形式。中央面积最大的圆柱体容纳通高的中庭、交通辅助空间和内部办公用房，周围的5个则是各类办事大厅。这使得中庭通往各办事大厅的空间识别性好，交通距离短，发挥了"一站式"服务的特点。建筑的主体结构采用大跨度多筒体体系和空心楼板系统。

To serve the key point of the building, a service plaza for citizens, its design is committed to openness, ample, facilitation and efficiency. There are six cylinders. While the largest one houses an atrium, traffic and service facilities and individual office cells, the other five are open spaces for approval service halls. It forms a high-efficient circulation system and well-defined place identification. To reinforce the architectural arrangement, the structural design adopts a large-span tube structure system and hollow floor slabs.

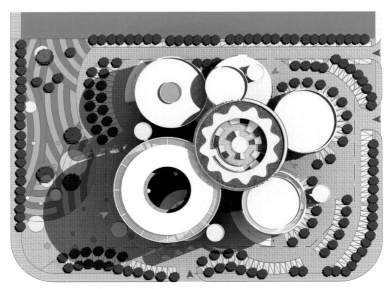

总平面图

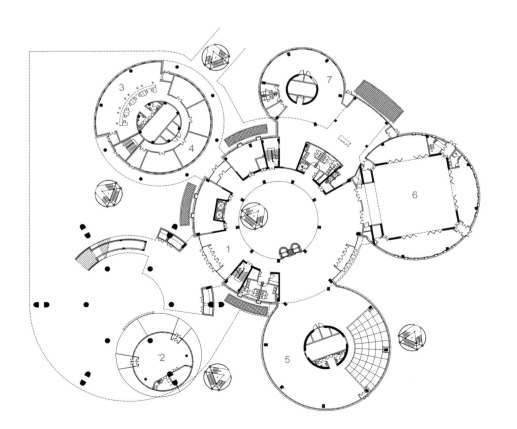

1. 门厅
2. 咖啡厅
3. 信访办
4. 办公
5. 城市展厅
6. 多功能厅
7. 厨房

首层平面图

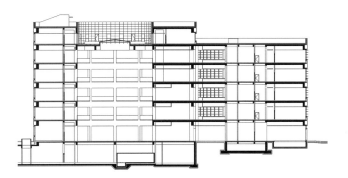
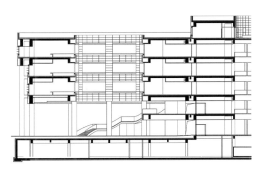

剖面图

197

镇江新区金融大厦 Financial Building of Zhenjiang New District

地点 江苏省镇江市 / 用地面积 15 500m² / 建筑面积 78 394m² / 高度 99m / 设计时间 2009年 / 建成时间 2012年

方案设计 崔海东　王敬先
设计主持 崔海东
建　　筑 王敬先
结　　构 张付奎　刘新国
给 排 水 杨兰兰　商　诚
设　　备 金　健
电　　气 马霄鹏
总　　图 刘晓琳
室　　内 邓雪映
景　　观 李　力

金融大厦是镇江新区核心区的第一个高层项目。主塔楼位于用地东北侧，裙房沿三侧展开，最大限度利用城市主要道路的沿街面，方便银行对外营业。塔楼底部三层布置有大堂、商务配套和休闲功能，与塔楼相连的裙房则为两家银行的营业厅，拥有各自独立的公众入口以及内部入口。稳重的立面形象符合金融业特点，塔楼的两个L形框架勾勒出结实稳重的体型。方形切角布局既退让出街角，避免对道路交叉口的影响，又与路东同期设计的酒店形成了环抱式的形态，像新区的大门向北敞开。由西北向东南的轴线则指向了近旁视野效果突出的湖泊。

Holding a prominent place in the new district of Zhenjiang City, it's the first high-rise building of this area. The tower, which anchors on the northeast corner, contains a lobby, business and entertainment facilities on first three floors and offices on upper floors. The podium facing three streets offers enough individual public entrances for two bank business halls and a stock trading hall in it. Two inverted L shapes frame a solidity volume for the tower. Filleted on the northeast corner, it steps back with its neighbor, the hotel also designed by CAG, serving as a gateway of the new district, while the longitudinal axis of the tower directs a panoramic view to the adjacent lake.

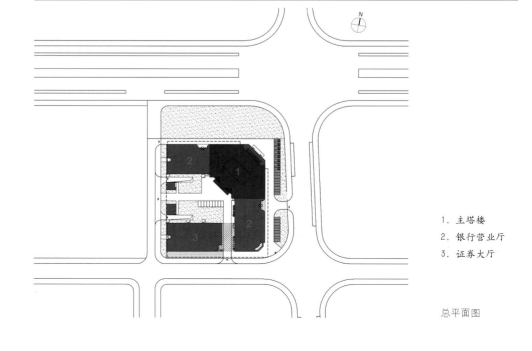

1. 主塔楼
2. 银行营业厅
3. 证券大厅

总平面图

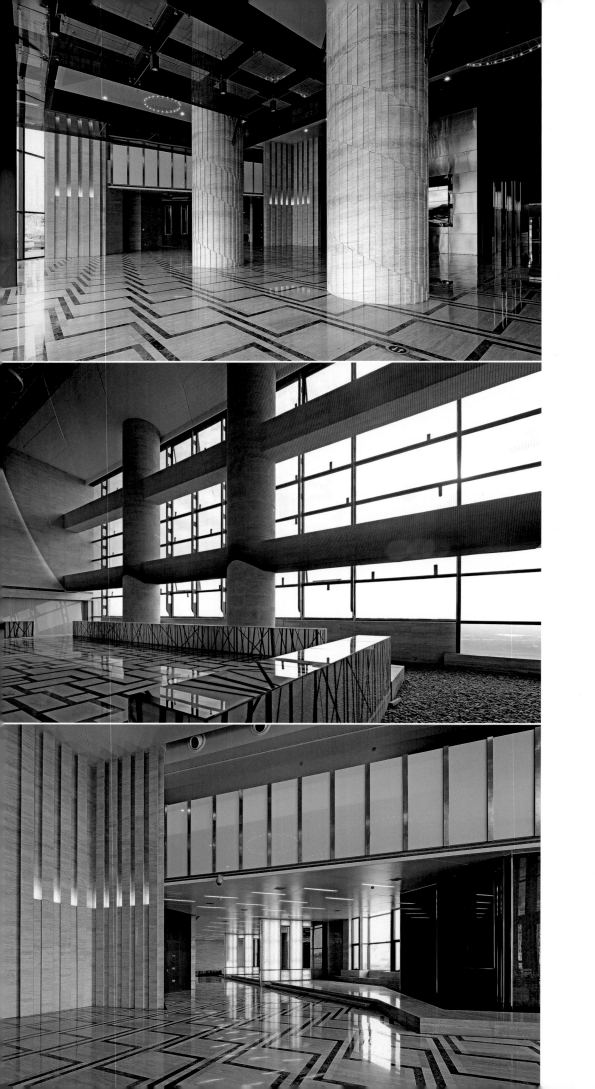

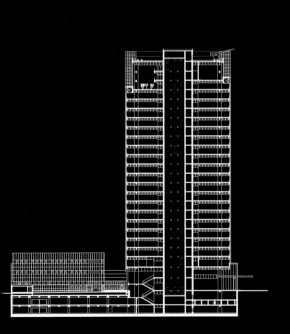

剖面图

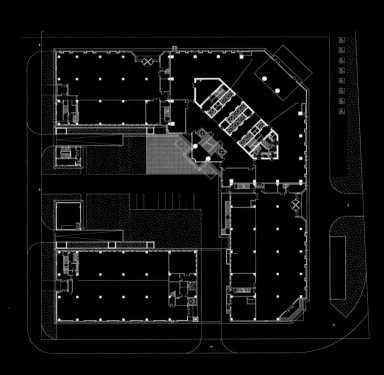

首层平面图

宁波市交通运输委员会办公楼 Office Building of Transport Committee of Ningbo Municipality

地点 浙江省宁波市 / 用地面积 18 600m² / 建筑面积 91 147m² / 高度 124m / 设计时间 2007年 / 建成时间 2010年

方案设计 徐磊　于海为　王宇航
设计主持 徐磊　王宇航
建　筑 汤小溪
结　构 王立波
给 排 水 张燕平
设　备 宋玫
电　气 王亚东
总　图 连荔

该项目为宁波东部新城的若干高层建筑之一。东部新城的规划采用小地块街区模式，并且严格规定了建筑的高度和外部街墙界面。两座塔楼分别用作交通局自用办公和商务办公，设计的难点是如何在逼仄的用地上布置十余个功能入口，形成互不干扰的流线，并符合人的行为感受。机动车流线按规划制定的方向组织到院落内部，解决了地库出入口和大堂落车问题。办公及商业入口都在塔楼外围解决。一致的框架形式，促成了对角布置的两座塔楼的对话。竖向条窗统一了塔楼和裙楼的肌理，通过疏密变化体现了两者差异。

As one of the office towers in the future CBD of Ningbo City, the building is designed within a strict urban planning of small block, defined height and complete street wall. Two office towers, one for the committee itself and one for rental, anchor the complex at two corners. The design challenge was to add more than 10 entrances into the narrow site without causing circulation interruption. By arranging the vehicle route into the inner court, most of the pedestrian entrances are arrayed outward. Apparently unified with the podium by the vertical stripes arranged in a rhythmical way, the towers remark themselves by the cantilevered structural frameworks indeed.

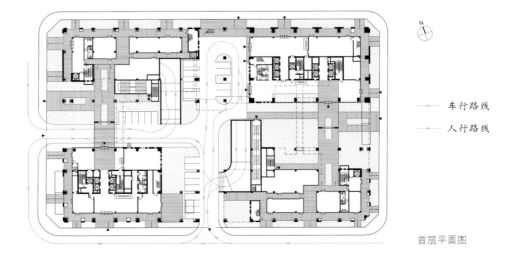

— 车行路线
— 人行路线

首层平面图

宁波市工商局办公楼　Office Building of Ningbo Administration for Industry & Commerce

地点 浙江省宁波市 / **用地面积** 7 159m² / **建筑面积** 44 584m² / **高度** 82m / **设计时间** 2008年 / **建成时间** 2012年

方案设计	徐　磊	汤小溪	刘　恒
设计主持	徐　磊	汤小溪	
建　筑	刘　恒		
结　构	张守峰	何相宇	
给排水	杨东辉		
设　备	宋　玫		
电　气	李俊民		
总　图	连　荔		

工商局办公楼坐落于交通委员会办公楼西侧，同样由18层自用办公楼、5层商务办公楼和商业裙房围合出完整的城市界面。群体整合的布局方式优化了功能关系和交通系统，是对高密度、多功能综合体需求的回应。底层部分架空，一体化设计的内部公共空间和院落、广场，形成视线的贯通及室内外空间的整体性。外观设计力求突出政府建筑的简洁大方，浑厚的陶瓷板框架与玻璃幕墙形成对比。立面以竖向线条为主，简洁的现代建筑语言与古典的比例划分相结合，增加了挺拔感和庄重感。由于主要朝向偏西，设计使实墙与开窗的比例接近，以节省能源。

Seated to the west of Transport Committee's building, this building follows the former's massing strategy to respond to the high-density and multi-uses demands for the urban complex. Besides the complete urban interface and the highly integrated circulation system, opening of the ground level is an innovative idea to allow views from the street to the inner court and integrate the design of exterior form, interior space and landscape. Still employing a vertical vocabulary, the ceramic plate clad stripes claim a dignified sense by adopting the classical façade proportion. Since the main façades are facing west, the area of glazed openings are minimized as large as the solid walls to reduce energy consumption.

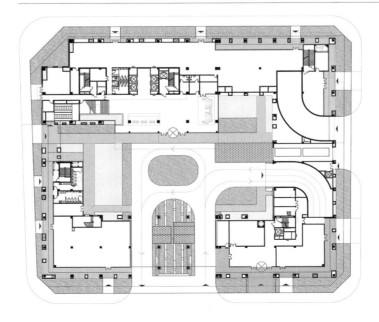

首层平面图

玉树康巴风情商街 Yushu Khamba Style Commercial Street

地点 青海省玉树县 / **用地面积** 112 200m² / **建筑面积** 71 000m² / **高度** 22m / **设计时间** 2012年 / **建成时间** 2013年

方案设计	刘燕辉 宋 波 李立宇
	李亚杰 李 翠 冯 涛
设计主持	刘燕辉 宋 波
建 筑	李立宇
结 构	董明海
给 排 水	陈 超
设 备	张 昕
电 气	王京生
总 图	邵守团
景 观	刘 环
合作设计	中国城市规划设计研究院
摄 影	刘燕辉 宋 波

本项目是玉树地震后的重建项目，由康巴风情商街和红卫滨水区两部分组成。康巴风情商街是一条连续的商街，整体性比较强，设计时贯彻了整合、融入的设计理念，将商业部分即首层和二层进行了整合设计，采用相同的建筑风格，而三层以上的住宅部分面积不同，体量各异，融入了多种建筑元素，形成下部完整上部丰富的外观特点。红卫滨水区则为前商后住，商业与住宅的建筑形式有明显区别，且各地块相对独立，所以在设计时采用了多样的原则，每个地块都带有各自独立的建筑特色。

As a reconstruction project after the Yushu Earthquake, the project contains two parts: the Khamba Style Commercial Street and the Hongwei Waterfront Zone. The commercial street carries on the continuous street space as before. The first and second floors are designed with same architecture vocabulary to form a unified commercial atmosphere, while the upper floors can be designed by various manners to show their private nature. For the modules of Hongwei zone, the residential rooms are added behind the store, so that multiple styles could be applied on them. So the appearances of those plots are much more diversified.

总平面图

项目位于两条河流的交汇处，是县城独具商业价值的滨水核心地区，也是向国内外游客和市民展示灾后重建水准、新玉树康巴特色风情的核心区域，定位为集商业、文化、旅游、休闲、居住为一体的综合性区域。

Located in the junction of two rivers, the commercial street occupies a pivotal point of Jiegu Town. As a complex community of retail, tourist service, entertainment and residence. Functions, it is the key area for the county seat of Yushu to show the reconstruction achievement and Khamba culture for the visitors from all over the world.

常州方圆·云山诗意　Changzhou Fineland · Orient Residential Area

地点 江苏省常州市　/　用地面积 72 600m²　/　建筑面积 240 000m²　/　高度 98m　/　设计时间 2008年　/　建成时间 2010年

设计机构　深圳华森建筑与工程设计顾问有限公司

"云山诗意"作为华森与方圆房地产合作的系列产品，其中江南民居形式与现代高层住宅的结合，以及在景观中丰富的传统意境，已具有独特的品牌效应。这一社区位于市中心周边，其中点式高层设置于用地中心，板式小高层则布置于南北道路两侧，通过错落有致的布局形成丰富的天际线。住宅通过现代材料和方式对古典的色彩、光影、雕饰形式进行了演绎。多层次的庭院空间，亭台榭舫的点缀，以及细致的景观设计，以景观园林的手段进一步增强了小区的东方气质。

The "Orient" Community is already a mature brand produced by the Fineland and HSA, a branch of CAG. The blending of traditional South-China architectural language and the contemporary high-rise residence, and the lushly landscaped gardens are both attractive qualities of the series. In this project, the residential towers are located in the center as well as the slab buildings are arrayed on the street sides. So that it creates a picturesque undulating skyline and enriches the street spaces. The appearance combines traditional tastes and modern styles by indigenous colors, materials, light and shadow.

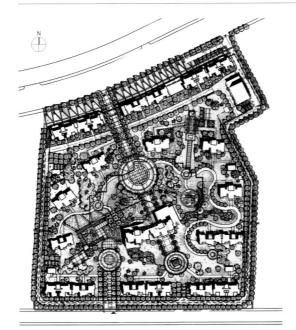

总平面图

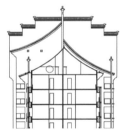

马头墙、坡屋顶、白墙、黛瓦等传统元素与传统江南民居取得联系

The classical elements, such as the pitched roof, white wall and black tile, show respects for the traditional buildings in South China.

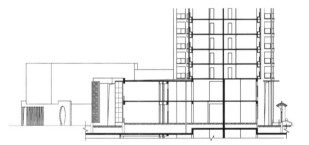

住宅楼前多层次的庭院空间和细腻的入口设计

Layering of garden spaces and the well-designed entry sequence

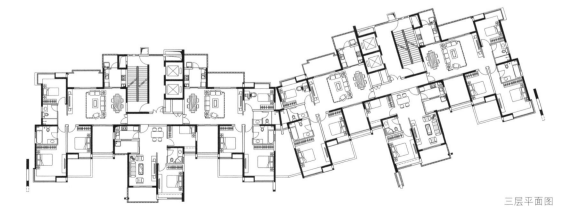

三层平面图

四合上院 Sihe Up Courtyard

地点 北京市西城区 / 用地面积 38 910m² / 建筑面积 199 260m² / 高度 45m / 设计时间 2008年 / 建成时间 2011年

方案设计	陈一峰　于 洁　尚 佳
	李慧琴　王春雷　段晓莉
设计主持	陈一峰　于 洁
建　筑	尚 佳　李慧琴　王春雷
	段晓莉　时 壮
结　构	王 奇　刘 巍
给 排 水	高 峰　贾 鑫
设　备	李雯筠　王 肃
电　气	贾京花
电　讯	凌 劼
总　图	余晓东

四合上院位于北京二环路内富有历史文脉的老城区，历史文化核心保护区琉璃厂的西侧。根据控制性规划要求，这里的建筑高度不得超过45m，但寸土寸金的地理位置又需要实现较高的容积率。小区规划沿袭了传统四合院内向围合的布局形式，采用12～15层的板式住宅，确保居住环境的舒适和通透，并实现了高达4.05的容积率。传统民居的灰墙瓦顶，通过适当的转译和恰当的尺度把握，与周围的城市风貌相融合。

West to the core zone of historical conservation protection of old Beijing, Liulichang, the dilemma of the design is to conform the zoning code without overcoming the height limitation, and to accommodate more residences in such a prime location at the same time. The layout continues the introverted plan of traditional courtyard with multilevel buildings. Not only gains a comfortable residential environment with natural light and ventilation, it also gain a FAR larger than 4. The grey brick and tile-coated pitched roof vocabulary allude to the nearby courtyards.

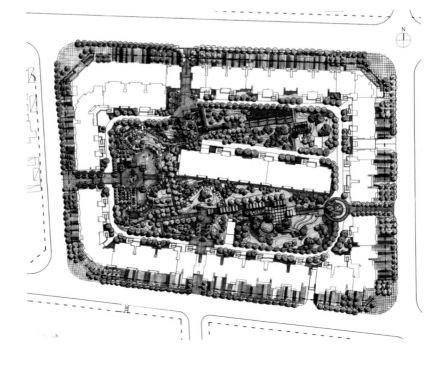

总平面图

重庆线外SOHO及会所　The Forward - SOHO & Club, Chongqing

地点 重庆市北部新区　/　用地面积 27 300m²　/　建筑面积 59 626m²　/　高度 99m　/　设计时间 2007年　/　建成时间 2010年

方案设计	陈一峰　王春雷
设计主持	陈一峰
建　　筑	王春雷　周　毅　陈旺军 黎　靓　王亦俊　林　琢
结　　构	李　华　熊　宇　吴晶晶
给 排 水	王学劼　娄和洲
设　　备	谭　涛
电　　气	陈德明
电　　讯	陈子军
图片提供	陈一峰

作为信息时代和网络时代的建筑，设计试图体现便捷、高效、互动的网络化理念。SOHO公寓楼被定义为垂直连续扭转体，东立面和西立面上的墙面扭转方向相反，结合水平的墙面形成了整个建筑的不规则骨架，转角处的幕墙竖肋更加突出了建筑的错动感。位于小区中心的会所，前期作为售楼处使用，以几个不同大小的虚实体块组成，辅以由电路板抽象而来的图案，夜晚由立面图案缝隙流淌下来的灯光一直延伸到水池池底，形成带有动感的光影效果。

The SOHO apartments and the club draw their design theme from the interaction and high-efficiency of the information and network era of this time, which is also the breeding bed of the SOHO living style. By inverting the place of solid and transparent glass curtain wall, the different units of the façade form a winding frame for the towers. The grilling on the corner also strengthens the dynamic sense. For the club, a series of solid or translucent cubes covered by the pattern abstracted from the texture of circuit board, are reflected by the water surface in front of them. At night, the lighting in the gaps flowing to the water intensifies the dramatic effects of light and shadow.

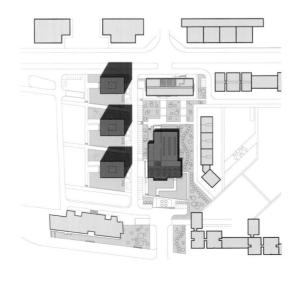

1. 会所
2. SOHO公寓楼

总平面图

北京风景 Beijing Scenery

地点 北京市昌平区 / 用地面积 168 803m² / 建筑面积 370 158m² / 高度 30m / 设计时间 2010年 / 建成时间 2012年

方案设计	陈一峰 王春雷
	尚 佳 张黎黎
设计主持	陈一峰
建　　筑	尚 佳 王春雷 张黎黎
结　　构	张淮湧 王 婧
给 排 水	陈 宁 赵 昕
设　　备	何海亮
电　　气	王玉卿
总　　图	王 炜

北京风景定位为宜居、舒适、绿色的住宅项目。整个园区以环形绿化带作为景观走廊，通过景观步行系统组织室外步行和活动场地，提高室外公共设施水平。设计采用强化院落，弱化组团的手法，数个可容纳100~200户的院落并联在小区干道两侧。绿化则由中心绿地、休闲步行场所和院落绿化三部分组成。小区采用外环式道路，从入口处开始人车分流，为居住者提供安全、自由的活动空间。立面设计采用简洁而严谨的现代建筑语言。以完整的手法将通常细碎的住宅立面整合为具有高档公寓和公共建筑性格的沉稳大气的形式。

The extensive landscaped community uses a landscape loop, which is composed of pedestrian route and activity space, linking four clusters of residential buildings. Although most of them are arranged in rank, the buildings also give a sense of court by emphasizing the landscape and some of irregular disposition. The greening system is formed by the three parts, the central garden, the pedestrian area and the courtyard. To present a bolder even like the public building gesture, the exterior reforms the conventional too sophisticated style of residential building into a crisp, modern manner which indicates the high-end of the buildings.

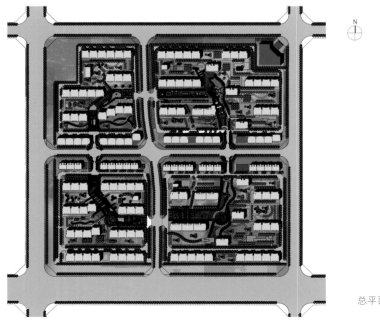

总平面图

标准层平面图

远洋一方二期　Poetry of River, Phase II

地点 北京市朝阳区 / **用地面积** 保障房 96 616m² 商品房 164 900m² / **建筑面积** 保障房 232 368m² 商品房 262 000m² / **高度** 60m
设计时间 2008年 / **建成时间** 2012年

方案设计	赵钿　王凌云　陈敬思　潘磊　张燕鹏
设计主持	赵钿　王凌云　古云
建　筑	李楠　杨文斌　陈敬思　潘磊　陈霞　陈敬思　韩风磊　张燕鹏
结　构	余蕾　孙洪波　何相宇　贾开　王玮　郑红卫　张守峰　段永飞　程颖杰　郝国龙
给排水	吴连荣　朱跃云
设　备	祝秀娟　刘筱屏
电　气	马霄鹏　王铮　何静
总　图	余晓东

远洋一方居住区位于城市周边地带，其二期建设由两部分组成，东为中小户型的中高档社区；西侧为配建的两限房和廉租房等保障性住房。保障性住房北邻通惠河，西有100m宽的绿化隔离带。规划充分利用这些景观资源，沿河布置高层板塔式住宅，沿西侧景观带则布置一梯三户的小户型。廉租房则靠近主要道路设置，便于管理。立面设计经过合理分析，避免不必要的造型因素增加成本。基本立面元素的集中和组合，使社区形象在统一中有所变化。保障房住宅均按板式或板点式进行设计，确保充足的采光通风条件，在有限面积内强调舒适性和经济性。

Situated on the outskirts of Beijing, the project contains two zones, the east one houses saleable residences and the west one is a community for affordable housing. The mast plan of the affordable zone sets the high-rise buildings on the north and the L-shaped buildings for small units on the west to maximum number of families that could view the northern river and the western green belt. On the other hand, a series of low-rent housings are laid along main roads to facilitate management. Unnecessary exterior elements are diminished to emphasis the economical efficiency, while natural lighting and ventilation are drawn into the units to maintain a comfort environment within the limited area.

1. 保障房地块
2. 商品房地块
3. 幼儿园
4. 托老所
5. 配套商业

总平面图

保障房一梯两户户型平面图

保障房一梯四户户型平面图

东侧商品房地块主要面向首次置业的年轻人群。规划采用人车分流、景观车库、就近回家的理念，保障住户的安全与便利。采用多级绿化体系，设置集中绿地、组团绿地、宅前绿地等，为住区提供多系统的活动交流空间。针对中小户型的特点，单元设计强调公共交通核心布局的紧凑，减小公摊面积，户型布局则强调功能分区明确、动静分离、尺度适中、功能细化，提供了衣帽间、储物间、家电等具体的使用空间。建筑外观风格现代、典雅，与环境融合构筑温馨宜居的人居社区。

Most of the purchasers of the saleable residences are young white-collars. The separation of pedestrian and vehicles, the landscaped garbage and the short-access to home pay much attention on the safety and convenience they cared. With a compacted public traffic core, the layout of the unit itself clarifies the different functional areas and adds a number of small spaces with dedicated functions, such as the coat room, storage and equipment locations. Modestly designed exterior is quite fit for the simple and livable concept of the community.

商品房小高层标准层平面图

万科·蓝山　Vanke Beijing Hills

地点 北京市朝阳区 / 用地面积 60 185m² / 建筑面积 184 000m² / 高度 80m / 设计时间 2007年 / 建成时间 2011年

方案设计	赵　钿　王凌云　潘　磊
设计主持	赵　钿　王凌云
建　　筑	潘　磊　杨晓昕　叶冬青
	白　宇　张燕鹏
结　　构	王　玮　郑红卫
给 排 水	陈　宁　周　博
设　　备	张　斌
电　　气	郭玉欣　陈沛仁
总　　图	王　炜
合作设计	日本津岛设计事务所

设计在规整的用地内，将建筑作为当代生活的背景和容器，围合出公共性的大庭院。为减少对住区的影响，幼儿园被置于用地东南角，同时获得充足的自然采光。南北侧为80m的高层住宅，西侧是东西向的多层住宅，东侧则为会所。车库采用半地下的方式，获得一定的采光，并丰富了景观的竖向层次。庭院内则不设车行道，完全用作居民的休闲场所。一梯两户的户型采用前后通透的布局，分区明确，流线合理，并细化了功能区，具有较高的舒适性。建筑外观则追求城市化特征，简洁有序，用楼层的错落控制天际线的变化。

The master plan of this project is to enclose a public garden for all of the residences here with concise and regular buildings. Regarding the slab buildings as container and background to present the contemporary life, it puts the 80m-high buildings on the north and south sides of the site, while the east-west oriented slab is seated on the west side. To obtain amply light, the kindergarten is settled on the southeast corner. The semi-underground parking not only reduces the artificial lighting, but also enriches space effects of the landscaped garden. Without distractions of vehicles, the central garden contains plenty amenities for residences.

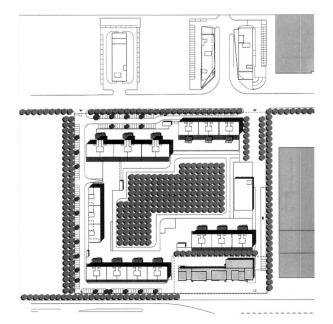

总平面图

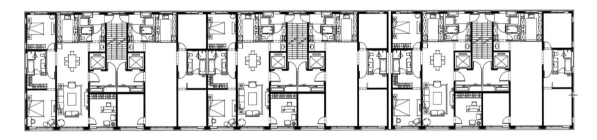

住宅标准层平面图

中海·苏黎世家 China Overseas · Zurich Garden

地点 北京市丰台区 / **用地面积** 22 300m² / **建筑面积** 68 936m² / **高度** 30m / **设计时间** 2010年 / **建成时间** 2013年

方案设计 张通 王越
设计主持 张通
建　筑 陈帅飞 王越
结　构 崔青 万小莉
给排水 袁乃荣 王松
设　备 刘燕军 郭然
电　气 贾京花 史敏
总　图 李可溯

本案位于公园环抱的城市外围地带，小区因而采用半围合式布局，西侧向城市绿地开放。南北沿红线放置两排10层高的单元式小高层住宅，形成小区外廓，东侧则以9层高的东西向住宅围合出小区入口，形成安静的内部庭院。住宅以小三居或小两居为主，满足套密度大于212套/ha的规划要求。小区道路分为内环和外环。外环为铺设植草砖的消防车专用道，内环则为人行路。内部庭院形成集中的公共绿化区域，通过尺度亲切的景观布置，为每个组团提供不同特点，以增强识别性和归属感。宅前小院既丰富了绿化层次，也为首层的居室提供了保证私密性的屏障。

Embraced by city parks, the project located in the outskirts of Beijing adopts a semi-closed layout to open to the western urban green belt. Two rows of high-rise slab buildings profile the site, while the main entrance is framed by two 9-story buildings on the east side. Most of the residences are two- and three-bedroom units, and the unit-density of the site is larger than 212/ha. The outer loop is a fire lane paved by grass brick, and the inner loop for pedestrian rings a central garden with small-scaled landscape. The small courts in front of buildings not only provide layering plants, but also screen views to the units on ground floor.

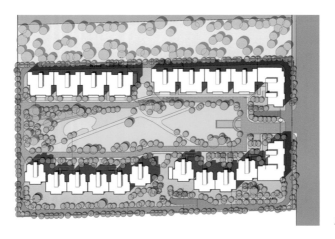 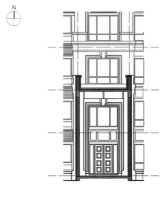

总平面图　　住宅入口详图

太湖湾云顶天海度假村 Taihuwan Yunding Tianhai Village

地点 江苏省常州市 / 用地面积 221 600m² / 建筑面积 110 627m² / 高度 25m / 设计时间 2008年 / 建成时间 2010年

设计机构　深圳华森建筑与工程设计顾问有限公司

整个度假村依水而建，由临水公建区和住宅区组成。公建区位于用地西南侧主入口处，沿内湖环形布置，底层为会所和小型餐饮、商业，上部为公寓。住宅区则被水面划分为四个小岛和两个半岛，主要由2～3层的联排和双拼别墅组成。由于基地整体地势较低，需要考虑太湖的防洪特殊要求，并尽可能减少土方量，设计根据地形采取了不同的台地处理方式，部分建筑采用错层设计，保证了建筑与水面、景观的良好关系。公建沿内湖环形布置，公寓人流通过外侧的集中大堂进入上部公寓。公寓根据面积大小不同分为外廊式，单元式和独立复式三种，均能获得开阔的景观视野。

Located on a waterfront site not far away from Taihu-lake, the village includes public facilities and residences. The public area ringing a lagoon contains a club, restaurants and retails in the ground level and apartments in the upper levels. The two-story and three-story townhouses in the residential area are grouped together into six clusters. To minimize the volume of earthwork and keep an intimate relation between buildings and the water surface within the strict constraints of flood control, the site is transformed into various terraces and some of the buildings adopt staggered floors. According to the size, there are three types of apartment: gallery-type, unit-type and double-level-type. All of them have open views to the lagoon.

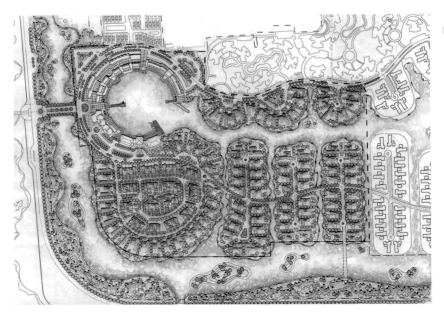

总平面图

兰州鸿运润园集成示范住宅 Integration Demonstrative Residence of Hongyun Run Garden, Lanzhou

地点 甘肃省兰州市 / **用地面积** 23 000m² / **建筑面积** 84 000m² / **高度** 60m / **设计时间** 2008年 / **建成时间** 2013年

方案设计	刘燕辉　韩亚非　胡璧
设计主持	刘燕辉　韩亚非
建　　筑	黄　路　师亚新
结　　构	张兰英　蔡玉龙
给 排 水	关　维
设　　备	张　昕
电　　气	王京生　张晓泉
总　　图	魏　曦
合作单位	日本市浦设计事务所

项目位于中国西部城市兰州，是中日技术集成住宅和健康住宅在我国西北地区的首次实践，采用套型优化与室内精装修的协同设计方法，对岛式卫生间、收纳空间、同层排水、厨房水平直排烟、地板采暖、全热式新风系统、太阳能与建筑一体化、高层分户太阳能热水系统等方面进行了探索。设计通过技术的引进和改良，建立了适合国情的技术集成应用住宅体系，在现有的传统施工技术下，采用精细化建造的手段进行本土化落地，以住宅产业化的方式在工厂定制保证体系和快捷的室内成套产品安装，为引进技术再创新积累了经验。

Located in Lanzhou, it is the first practice of China-Japan technology integration residence in northwest China. By the optimization of units and the collaborative design, it makes innovative exploration in land-style bathroom, storage space, same floor drainage and smoke extraction, floor heating, total heat fresh air system, solar panel integration and solar heating water. The remoulded technologies are much more adopted for the Chinese context that could employ the common building methods to realize fine construction in local projects and gain experience for the future re-innovation of introduced technology.

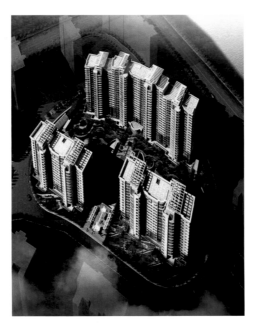

鸟瞰透视图

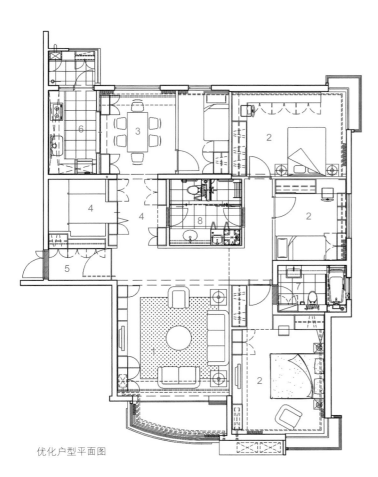

优化户型平面图

原始户型平面图

户型优化目标

· 优化结构承重墙体，减少原外墙过多的凹凸面
· 进门设玄关和对景墙
· 餐厅和客厅形成良好的对位关系，视线通透
· 边单元户型利用端部优势，实现两个方向的采光
· 根据中餐饮食习惯设炉灶，与餐厅采用灵活隔断
· 部分户型考虑做成餐厨合一，便于个性化选择
· 南北相对的居室门避免出现错位现象
· 居室门、窗关系避免通风死角

1、2 家具均采用预先定制、工厂化的生产方式，降低了装修成本，也确保了精装修房屋的施工精度和产品质量，保证室内空气质量

3 餐厅和客厅形成良好的对位关系，视线通透

4 结合室内精装修，见缝插针增加储物空间，分类明确

5 入户设置独立玄关，更衣柜要满足长短衣物收纳需求，鞋类洁污分离存放，既增强私密性也完善了住宅功能

6 采用工厂定制的整体厨房，按"洗、切、烧"流程布置，并保证一定的操作台长度，洗池、炉灶、烤箱、微波炉与冰箱等设备结合橱柜一体化整合设计

7 卫生间做到卫浴分离，采用同层排水，卫生洁具可灵活布置，同时便于检修，减少渗漏

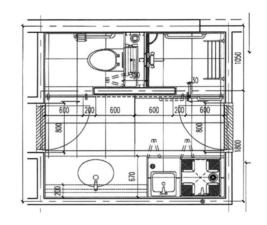

岛式卫生间平面图

8 岛式卫生间两端设门，便于不同室内空间就近使用。为满足共用需求，洁具独立分间方式，坐便、淋浴有独立隔间。洗衣机结合储物柜定位布置，设专用水龙头、托盘及地漏

精装修结合家具设计，将墙垛适当加大，实现了各种储物家具、电视柜、空调设施的设置与新风管道的隐藏，多种功能经整合形成家具墙面，大大增加室内空间的完整性

立面飘窗下和空调格栅外设有太阳能集热器,通过各户内的储水箱实现热水供给。在小区前期分户燃气壁挂炉采暖及生活供热水系统的基础上,仅增加分户太阳能循环和控制系统,即可实现太阳能的充分、有效利用

全热式新风系统同时强制将新鲜空气送入室内,并将室内污浊空气排出室外,以保持良好的通风

地下车库内设置自然采光窗井,减少人工照明,降低了日常运营成本

厨房烟道采用独立水平直排方式,避免产生各户串味问题

复地朗香别墅四期 Forte Ronchamps Villa, Phase IV

地点 江苏省南京市 / 用地面积 140 000m² / 建筑面积 70 000m² / 高度 12m / 设计时间 2009年 / 建成时间 2013年

设计机构 深圳华森建筑与工程设计顾问有限公司

复地朗香别墅区位于毗邻湖畔的山麓南侧，基地南面与一期用地相接，设置了三排侧院式独栋别墅，屏蔽道路对基地的影响。沿坡度向上则布置全院式独栋别墅，户型面积随高度及景观视线的提高沿坡度上升，在最北侧放置三层高的大全院式独栋别墅，尽享山水之景。内部主要道路为5m单行道，形成环道，方便管理，也保证了小区的私密性。环境设计利用不同标高的绿化空间，营造富有层次感的立体园景。户型设计创新地采用了侧院式独栋别墅，别墅的主要功能房间均能享受到院落景观，增强了院落的使用率。

The villas occupy a hillside overlooking the scenic lake and bordering the first phase of the project. Three rows of medium-scale villas are settled on the side of main entry, while larger villas are settled on the slope to provide a better view of the lake. On the top of the site are the largest villas with expansive garden, which gesture to the valley beyond. To gain a quite atmosphere, the community adopts 5m-wide one-way loop to satisfy the traffic demands. The medium-scale villa with a garden beside its gable wall is a creative prototype to improve living quality in limited area. The windows of main rooms are opened to the gardens to maximize views for them.

总平面图

小区整体剖面图

南京苏宁睿城　Nanjing Suning Smart City

地点 江苏省南京市　/　**用地面积** 49 300m²　/　**建筑面积** 194 000m²　/　**高度** 100m　/　**设计时间** 2008年　/　**建成时间** 2011年

设计机构　深圳华森建筑与工程设计顾问有限公司

小区的布局呈南低北高的趋势。南面沿街为18层住宅以及四层的社区中心，减轻高大体量对城市的压迫感，也易于产生丰富的立面轮廓线。外围及北侧则布置100m高的板楼。小区开放与围合兼备的独特格局强化了空间的领域感，在地块中部通过四栋建筑的底层架空，形成了南北200m，东西65m的大型中央景观，作为整个地块的骨架。围绕中央景观布置的住宅均采用南北通厅或端头主卧套的户型，尽量使住户都能享有良好的景观。住宅生活区则自成组团，与绿化轴紧密结合。

For such a site with a height-difference that the north is higher than the south, the 100m high slab type buildings are set in the north, while lower buildings and the 4-story community center are situated in the middle or the south. A central landscape which is more than 200m long and 65m wide is the kernel of the community. Such a spacious garden is created by the design strategy to open the bottom levels of four surrounding buildings. By set the living rooms and master bedroom facing the landscape, most of the units could receive a great view.

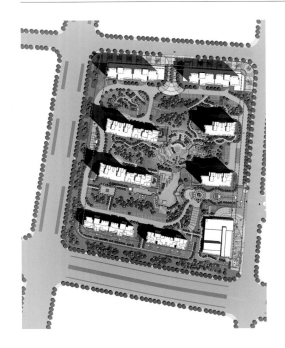

总平面图

大同市妇女儿童医院　Datong Women's & Children's Hospital

地点 山西省大同市 / **用地面积** 169 899m² / **建筑面积** 118 076m² / **高度** 50m / **设计时间** 2009年 / **建成时间** 2012年

方案设计	陈一峰　赵　强　杨　光　崔　磊　崔博森
设计主持	陈一峰　赵　强
建　筑	崔博森　杨　光　崔　磊
结　构	张亚东　尤天直　潘敏华
给排水	王则慧
设　备	徐　征
电　气	李战赠
总　图	连　荔

大同市妇女儿童医院综合楼主要由门急诊、住院、综合医技检查、健康体检和后勤办公等功能空间组成。各个功能部分通过宽敞明亮的连廊联系在一起，方便病患医护人员使用。由于创新性地把住院楼分为儿科、妇科、产科三个单元，使得就诊单元明确，功能互相结合得更加紧密。清晰的组织结构不但为人们提供了良好的可识别性，同时也为医院创造了富有特色的、舒适的就医环境。

The complex includes an outpatient building, three inpatient buildings, a medical technology building and other service facilities. To facilitate the circulation of medical staffs and patients, all of the functional areas are connected by several corridors with ample light and views. By dividing inpatient area into three units for pediatrics department, gynecology department and obstetrical department separately, the functional organization of each unit is well-defined and recognizable.

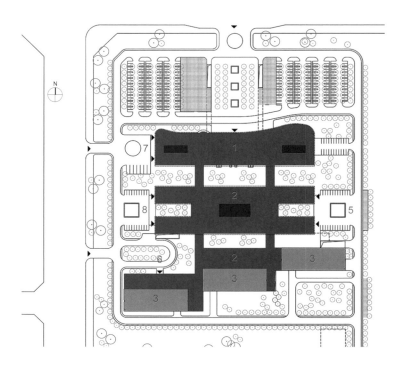

1. 门诊楼
2. 医技楼
3. 住院楼
4. 门诊入口
5. 办公入口
6. 住院入口
7. 急诊入口
8. 体检入口

总平面图

大同市中医院 Datong Chinese Medicine Hospital

地点 山西省大同市 / 用地面积 126 172m² / 建筑面积 88 890m² / 高度 51m / 设计时间 2009年 / 建成时间 2013年

方案设计	陈一峰 赵 强 杨 光 崔 磊 崔博森
设计主持	陈一峰 赵 强
建 筑	崔 磊 杨 光 崔博森
结 构	孙海林 霍文营 李 谦
给 排 水	王则慧
设 备	徐 征 刘玉春
电 气	王 琼 李战赠
总 图	连 荔

大同市中医院的设计吸收了山西传统民居院落的布局特点，按照一定的秩序，将庭院串联组合在一起。经过简化的院落空间，与建筑虚实相间，形成具有一定韵律变化的环境。建筑造型同样吸收了传统民居单坡屋顶的形式，屋面内倾，强调内部良好的院落环境。设计同时引用"经络"这一中医治疗的独特之处，蜿蜒曲折的坡屋顶形成贯穿整体的建筑"经络"，暗合中医学的经络理念。

According to the layout of the traditional courtyard in Shanxi, a series of courts are inserted into the plan to form the interweaving space system of interior and exterior. The single pitch roof, another famous feature of local house, is also used to visualize the spine of the hospital. The spine not only connects most of the wards as the circulation of the hospital, but also implies the "meridian" theory of Chinese Medicine.

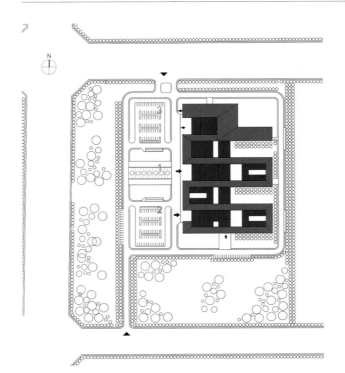

1. 门诊入口
2. 急诊入口
3. 住院入口

总平面图

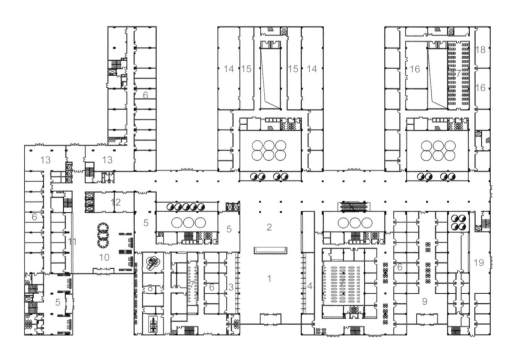

1. 门诊大厅
2. 接待
3. 收费挂号
4. 药房
5. 候诊
6. 诊室
7. 输液
8. X光
9. 急诊大厅
10. 住院大厅
11. 住院手续
12. 商店
13. 体检大厅
14. 制液配置
15. 摆药间
16. 阅览室
17. 图书馆
18. 电子病历中心
19. 餐厅

首层平面图

山西省肿瘤医院放疗医技综合楼 Comprehensive Building of Shanxi Provincial Tumor Hospital

地点 山西省太原市 / 用地面积 9 500m² / 建筑面积 52 509m² / 高度 79m / 设计时间 2009年 / 建成时间 2012年

方案设计　陈一峰　赵　强　杨　光
　　　　　崔　磊　崔博森
设计主持　陈一峰　赵　强
建　　筑　崔　磊　杨　光　崔博森
结　　构　王树乐　张淮湧
给 排 水　王则慧
设　　备　刘玉春　徐　征
电　　气　李战赠

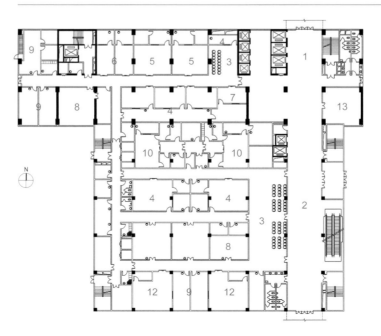

1. 门厅
2. 医院街
3. 候诊
4. 登记处
5. 数字化胃肠机
6. X线诊断
7. 乳腺机
8. 阅片室
9. 办公室
10. 血管造影像检查
11. 示教室
12. 磁共振影像检查
13. 商店

首层平面图

医院街

候诊空间

手术室

检验中心病理科

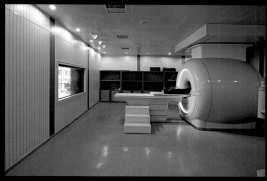
核磁共振

超声中心

护士站

单间病房

山西省肿瘤医院放疗医技综合楼的设计，从院区整体规划入手，重新完善肿瘤医院的医疗流程，既满足现代化医院的使用需求，又要能够满足医院不停诊、不停运行的要求。通过统一长远的整体规划，切实可行的分期实施，使得肿瘤医院放疗楼的单体设计自然和谐地融入肿瘤医院的空间秩序和人文文脉。在医疗流程方面，也使得原有相对混乱和交叉的各种流线得以重新梳理。内部医疗流线组织基于"医院街"的空间架构展开，科学地解决好人流、物流的组织，各功能区之间保证全天候、无障碍通行。

The construction of the radiotherapy technology building is considered as a redevelop of Shanxi Provincial Tumor Hospital on its campus to improve the medical process of this modern healthcare facility without interrupting the daily operations. The long-term strategy and stage planning make the building be a part of the whole campus naturally. The rearranged circulation system avoids its former chaos and interaction, making the curative facilities developing along the "hospital street system" to separate the activities of patients and logistics and guarantee the all time operation of every function areas.

欧美同学会改扩建　Extension of Western Returned Scholars Association

地点 北京市东城区　／　**用地面积** 900m² ／ **建筑面积** 2 850m² ／ **高度** 8m ／ **设计时间** 2009年 ／ **建成时间** 2013年

方案设计	崔　愷　傅晓铭　王可尧
设计主持	崔　愷
建　　筑	傅晓铭　单立欣
	刘　恒　冯　君
结　　构	王　载　刘　福　陈　明
给 排 水	尹腾文
设　　备	姜海元
电　　气	程培新　刘　畅
总　　图	高　治
室　　内	邓雪映

欧美同学会位于北京皇城保护区的核心位置，毗邻故宫，其院内的普胜寺为东城区文物保护单位。扩建工程需要在满足办公、会议和多功能活动厅等若干功能的同时，注重保护和改善周边古建筑的环境，探索古城保护和有机更新的途径。改扩建设计充分利用有限空间，通过向地下发展，满足大空间活动的功能要求。建筑造型既结合传统坡屋顶建筑特征，又使用现代建筑语言。严格控制的高度，使建筑没有对文物建筑和周边环境造成视觉影响。灰色金属隔栅形成的坡屋顶，将建筑体量减小，并与周边建筑自然地融合。屋顶平台的景观设计，则使建筑第五立面得以美化，改善了从北京饭店贵宾楼看皇城的景观。

The campus of the association is located in a district of historic center of Beijing, two blocks from the Forbidden City. The extension reconciles the architecture of the protected Pusheng Temple and others built later in the campus and creates a place to accommodate a gym, meeting rooms and offices. In the extremely limited site, the extension develops to underground to satisfy the requirements of large space. Working within the heritage of classical architecture, the design juxtaposed the respect for tradition with a distinctly modern vocabulary of concrete walls and the pitched roofs covered with metal grilling which diminishes the massing to tie the new building to its neighbors. The carefully designed plat roof is a result to improve the view from the nearby hotel.

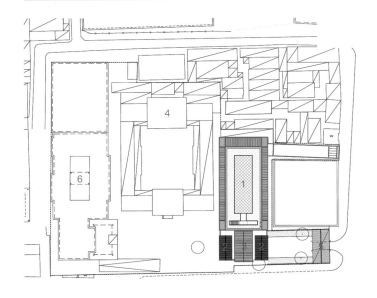

1. 新建建筑主体
2. 新建仿古门楼
3. 钢结构门廊
4. 普胜寺
5. 原有门楼
6. 原有建筑

总平面图

新增加了一进门楼，将原本不规则的院内空地划分为两进院落，不但对建筑主体有很好的遮挡作用，也提供了具有过渡性质的门廊空间。建筑造型既结合传统坡屋顶建筑特征，又使用现代建筑语言体现旧建筑改造的创新理念。

A pair of vestibules anchors the front court and creates a series of spatial process in the former irregular space. They not only screen the massing of the extension from the street, but also provide a transitional space between the entry and the protected historic temple. Both the original Chinese and modern vocabularies are used to echo to their historic neighbors and emphasis the modern nature of the extension.

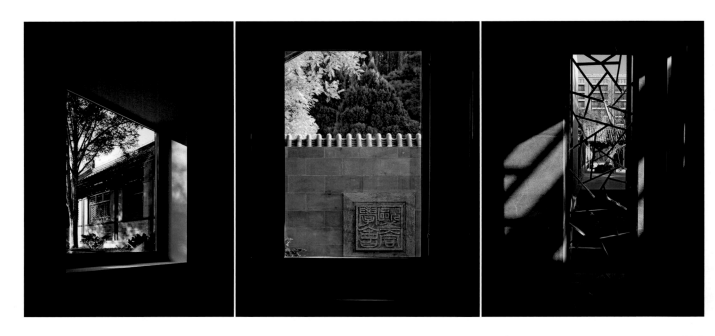

威克多制衣中心　Vicutu Garments Manufacturing Center

地点 北京市大兴区 / 用地面积 20 364m² / 建筑面积 310 454m² / 高度 28m / 设计时间 2010年 / 建成时间 2013年

方案设计　景　泉　李静威
　　　　　张文娟　肇灵犇
设计主持　李静威　杜　滨
建　　筑　张文娟　杜　捷
结　　构　王树乐　郭俊杰
给 排 水　黎　松
设　　备　梁　琳
电　　气　王苏阳
总　　图　高　治　刘晓琳
室　　内　张　晔　纪　岩

这是一家以生产高档男装为主的企业，需要将原有5层的框架结构生产车间改造为企业的运营、设计中心。设计充分发挥了原有框架的优势，在其顶部增加一层，以满足功能扩展需求；同时向南侧扩建，形成出挑的边庭，为建筑的主立面改造带来相对自由的条件。设计贯穿一体化的理念，实现了景观、建筑和室内的整体控制。改造后的建筑形体整合为一个轻盈自由的玻璃体和一个厚重的背景体块互相穿插的体量关系。原建筑东西向流线较长，设计在建筑中部增加了一部开放楼梯，穿针引线般联系起建筑的不同楼层，为办公空间提供了沟通交流的场所；出挑的平台实现了人们在空间中的互动，同时更加亲近景观，营造了一个丰富、宜人的边庭空间，二者结合而成整个办公楼的灵魂空间，有趣且人性化。

An enterprise focusing on top grade garments, Vicutu, needs to revise its five-story manufacturing factory into a management and design center. The renovation maintains most of the original structure and only adds one story on the top and extends on the south side to form an atrium, which is claded by a multi-layered curtain wall. The overall concept runs through the architecture, landscape and interior design. The revised structure becomes the interlocking of alight glass volume and a thick solid volume. The open stairs inserted into the middle shortens the east-west direction circulation as well as the cantilevered hexagonal balconies acts as the communication spots in the office space. Collaborating to be a spatial system, the new-add elements become the focal points of interior with dramatic effects.

原有5层的框架结构生产车间

边庭采用单层索网点驳接幕墙以实现较好的通透性,使室内环境与室外景观互相渗透,幕墙上下电动百叶的设置,充分满足了建筑的通风要求。背景主体采用陶板幕墙,材料本身具有节能环保的特点,同时呼应原有厂区的色彩基调,使整个园区得到统一与提升。

The taut curtain wall with high transparence offers an interpenetration of the indoor and outdoor spaces, while the ventilation requirements are satisfied by the power-driven shutters on the top and bottom of the façade. By cladding the solid volume with a kind of environmental friendly ceramic plate, the exterior wall palette reinforces the architectural cohesiveness of the surrounding buildings and improves the space quality of the whole factory relatively.

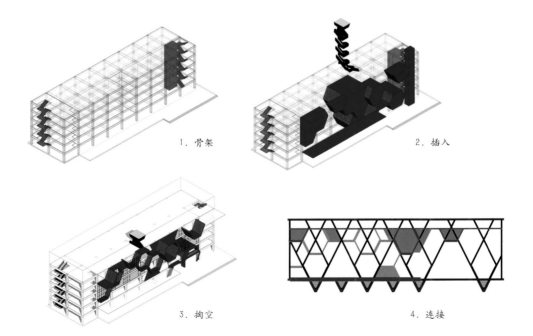

1. 骨架　　2. 插入　　3. 掏空　　4. 连接　　分析图

六边形盒子局部与幕墙结构体系连接,既解决了大跨度出挑的结构支撑,同时满足了内部空间的通风需要,并营造出更亲近景观的室内阳台。

Connected with the curtain wall structure, the cantilevered hexagonal modules act as ventilation tubes and a series of indoor balconies.

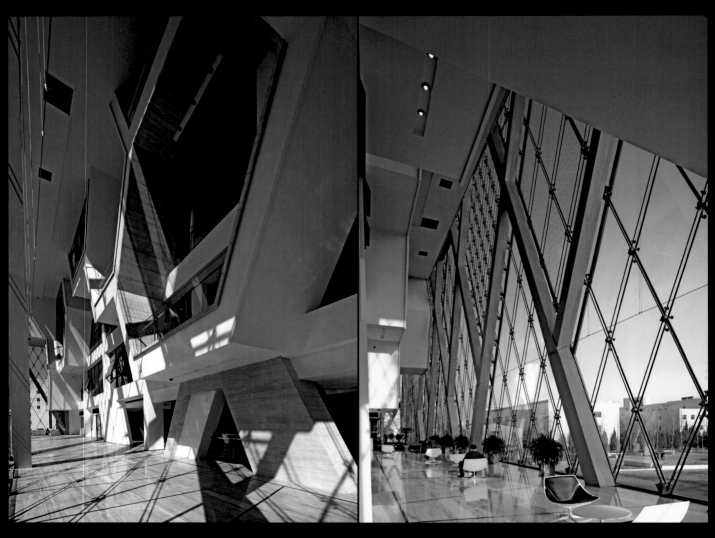
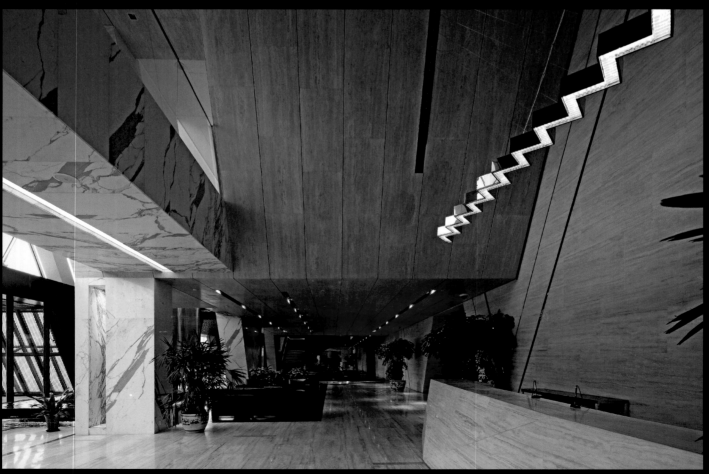

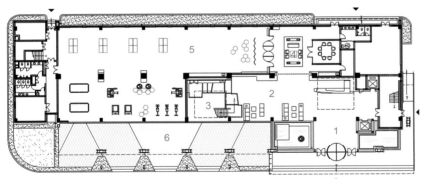

1. 门厅
2. 休息区
3. 企业文化展示区
4. 洽谈室
5. 多功能区
6. 时尚廊

首层平面图

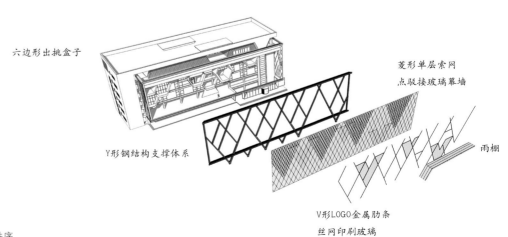

六边形出挑盒子

菱形单层索网

点驳接玻璃幕墙

Y形钢结构支撑体系

雨棚

V形LOGO金属肋条

丝网印刷玻璃

玻璃幕墙由外而内的秩序

中间建筑A、F区　A/F Plots of the Inside-out

地点 北京市海淀区　/　用地面积 28 243m²　/　建筑面积 36 579m²　/　高度 15m　/　设计时间 2008年　/　建成时间 2013年

方案设计　崔愷　喻弢　关飞　邓烨
设计主持　崔愷　时红
建　筑　喻弢　关飞
结　构　朱炳寅　张猛
给 排 水　靳晓红
设　备　杨向红
电　气　许士骅
总　图　连荔
室　内　张晔　韩文文

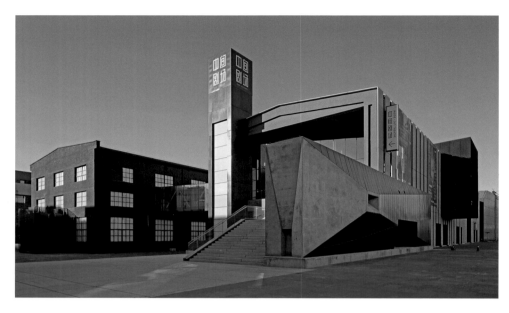

中间建筑位于北京西郊，是一处集居住、工作和艺术活动于一体的大型复合社区。A、F区为整个项目的最后一期建设，占据了社区面向城市主要道路的300m的城市界面。设计没有选择常规的综合体模式来解决多种功能需求，而是将功能单纯"形体化"，变成一个个具有独特外观识别性的房子，在沿街面上展开，错落的建筑之间形成了相互渗透和互动的城市空间。 其中A区为艺术街区，保留的厂房被改造为实验剧场，与厂房平行放置的艺术工作区则被有意表现为厂房剖面的复制，隐含了对旧日厂区的追忆。西侧的F区作为商业街坊，以"分户墙体"作为空间主导元素，墙体内集中交通和辅助功能，墙体之间交错布置商业空间和平台、院落。

The A/F Lots were designed to complement the whole Inside-out community, which is a compact village that accommodates diverse functions, such as residence, workshop and arts facilities. Instead of merging all the functions into a common "urban complex", its master plan visualizes the functions within individual and distinctive modules. Developing along the main urban street to the north, the seemingly scattered volumes create an activating space for urban activities. The art workshops, which stand beside a reformed factory, an experimental theatre now, refer to the factory's structural system to echo the original industry feature of the site. In the commercial street of F Lot, the gable walls between stores are thickened to house stairs and service space and become the dominant element that defines spaces and supports a series of platforms.

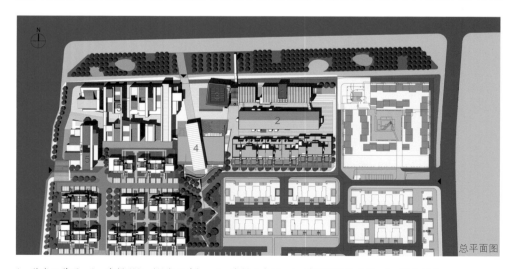

总平面图

1. 艺术工作区　2. 中间剧场（保留厂房）　3. 中间美术馆　4. 中间影院及会所　5. 商业街坊

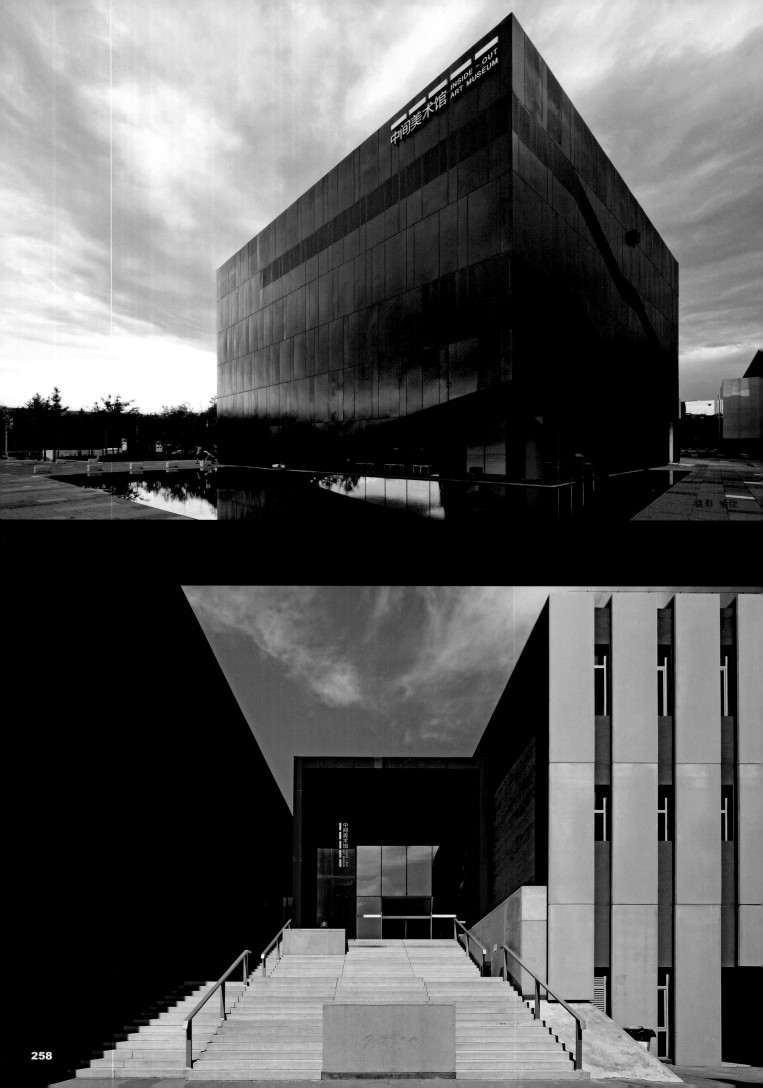

A black steel-clad "box", the art gallery, and the pure white tube are the focal points of the complex. Rotated in a subtle degree, they give the entrance plaza of the community a sense of place.

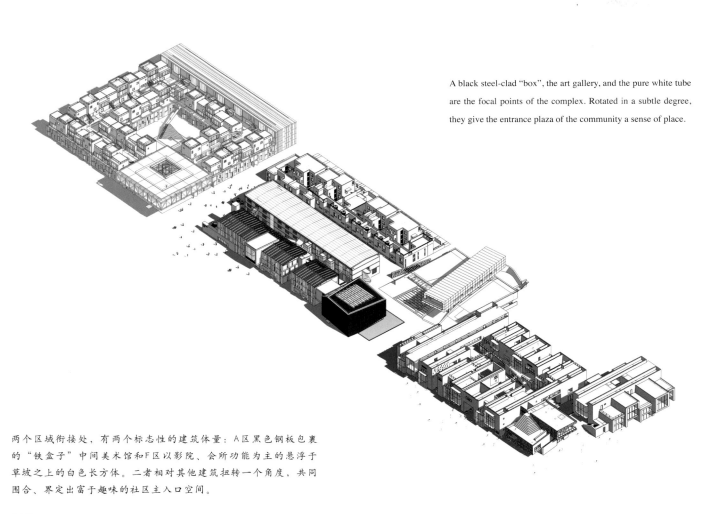

两个区域衔接处,有两个标志性的建筑体量:A区黑色钢板包裹的"铁盒子"中间美术馆和F区以影院、会所功能为主的悬浮于草坡之上的白色长方体。二者相对其他建筑扭转一个角度,共同围合、界定出富于趣味的社区主入口空间。

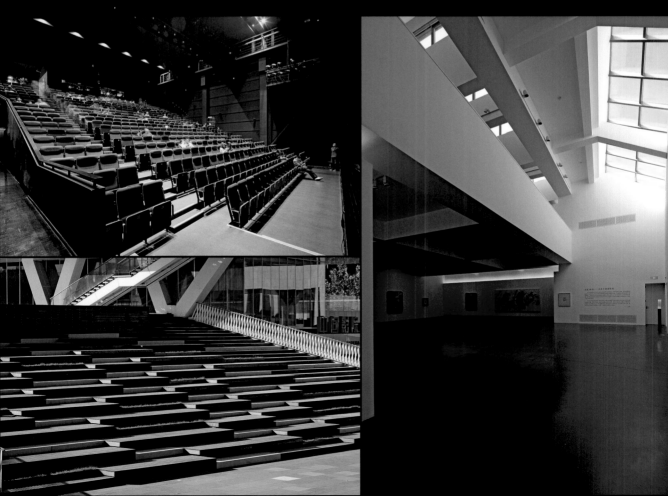

八达岭太阳能热发电站吸热塔 Thermal Absorbtion Tower of Badaling Solar Thermal Power Station

地点 北京市延庆县 / 用地面积 19 200m² / 建筑面积 2 536m² / 高度 119m / 设计时间 2008年 / 建成时间 2013年

方案设计　文　兵　张　波　白　晶
设计主持　文　兵　张　波
建　　筑　白　晶　刘　磊
结　　构　王　载　陈　明
给 排 水　董　超
设　　备　刘筱屏
电　　气　李建波　曹　磊

八达岭太阳能热发电站是"十一五"国家高技术研究发展计划的重要课题，也是国内首例太阳能热发电实验项目。工程整体包括试验基地和镜场两部分。吸热塔作为实验性构筑物，位于试验站中部，镜场南端。建筑造型来源于对太阳形象的抽象，通过体现太阳光芒的多边形平面不断旋转上升，形成流畅统一且极具视觉冲击力的整体形象。恰当的结构处理实现了空间扭转上升的复杂建筑造型。螺旋状的结构也形成富于动感的室内空间。嵌入外墙的采光窗使一层的公共空间如同布满星辰的夜空，在提供科普教育功能之余，也增添了引人探索的神秘氛围。

As a major subject of the "11th Five-year" National High Technology Research Plan and the first solar thermal power station in China, the station is composed of a test base and the mirror field. The thermal absorbtion tower is located in the middle of the site and south to the mirror field. The spiral derived from the polygon plan which is an abstraction of sun rays, is exactly a dramatic kernel point of the station. Adequate structure design methods realize the complicated spiral and twist construction. As well as the inner spaces, especially the first floor public space for popular science education, which is decorated by lots of day lighting holes, receive a fantastic atmosphere.

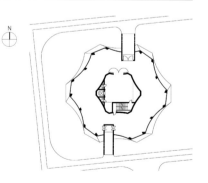

首层平面图

镜场内100面定日镜随着太阳升降转动，将阳光反射到吸热塔顶部3个吸热器，产生800℃的高温。

100 heliostats follow the daily path of the sun and reflect sunlight to absorbers on the top of the tower to heat water becoming 800℃ steam.

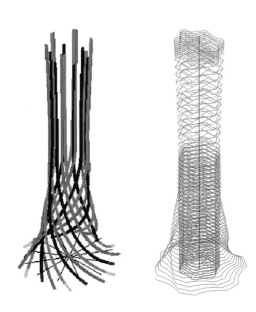

吸热塔外墙48.8m标高以下采用外裙内筒双层结构，以上部分内筒与外裙结构合二而一。内筒及塔身采用滑模工艺建造，外裙摆通过小模板拟合空间曲面。以每1.25m标高进行一次截面切片。

The structure under 48.8m is a double layer structure. Above 48.8m, two layers are combined together. The inner tube is constructed by slip form, while the space curved surface of the outer skirt is matched by small form boards.

图书在版编目（CIP）数据

作品2012-2013：中国建筑设计研究院作品选/中国建筑设计研究院编.
-北京：中国建筑工业出版社，2014.3
ISBN 978-7-112-16566-7

Ⅰ.①作… Ⅱ.①中… Ⅲ.①建筑设计－作品集－中国－现代 Ⅳ.①TU206

中国版本图书馆CIP数据核字（2014）第052622号

主　　编：崔　愷
执行主编：张广源　任　浩
美术编辑：徐乐乐
英文翻译：任　浩

建筑摄影：张广源

责任编辑：徐晓飞　张　明
责任校对：赵　颖

作品2012-2013
中国建筑设计研究院作品选
SELECTED WORKS 2012-2013
OF CHINA ARCHITECTURE DESIGN & RESEARCH GROUP

中国建筑设计研究院　编
*
中国建筑工业出版社出版、发行（北京西郊百万庄）
各地新华书店、建筑书店经销
北京雅昌彩色印刷有限公司印刷
*
开本：880×1230毫米　1/16　印张：16½　字数：420千字
2014年4月第一版　2014年4月第一次印刷
定价：**180.00**元
ISBN 978-7-112-16566-7
　　　（25414）

版权所有　翻印必究
如有印装质量问题，可寄本社退换
（邮政编码　100037）